Productive Bee-keeping
Modern Methods of Production and Marketing of Honey

by Frank C. Pellett

with an introduction by Jackson Chambers

This work contains material that was originally published in 1916.

This publication is within the Public Domain.

This edition is reprinted for educational purposes
and in accordance with all applicable Federal Laws.

Introduction Copyright 2018 by Jackson Chambers

COVER CREDITS

Front Cover
Beekeeper 2017 Honeybee Conservancy, College of DuPage by By COD Newsroom
[CC BY 2.0 - http://creativecommons.org/licenses/by/2.0],
via Wikimedia Commons

Back Cover
Mat Ong ABC by AmieKim (Own work)
[CC BY-SA 3.0 - https://creativecommons.org/licenses/by-sa/3.0],
via Wikimedia Commons

Research / Resources
The Top 6 Raw Honey Benefits
https://www.healthline.com/health/food-nutrition/top-raw-honey-benefits#1
via HealthLine.com

Wikimedia Commons
www.Commons.Wikimedia.org

Many thanks to all the incredible photographers, artists,
researchers, and archivists who share their great work.

PLEASE NOTE :
As with all reprinted books of this age that are intended to perfectly reproduce the original edition, considerable pains and effort had to be undertaken to correct fading and sometimes outright damage to existing proofs of this title. At times, this task can be quite monumental, requiring an almost total rebuilding of some pages from digital proofs of multiple copies. Despite this, imperfections still sometimes exist in the final proof and may detract slightly from the visual appearance of the text.

DISCLAIMER :
Due to the age of this book, some methods or practices may have been deemed unsafe or unacceptable in the interim years. In utilizing the information herein, you do so at your own risk. We republish antiquarian books without judgment or revisionism, solely for their historical and cultural importance, and for educational purposes.

Self Reliance Books

Get more historic titles on animal and stock breeding, gardening and old fashioned skills by visiting us at:

http://selfreliancebooks.blogspot.com/

introduction

Here at **Self-Reliance Books** we are dedicated to bringing you the best in *dusty-old-book-knowledge* to help you in your quest for self-sufficiency and food independence. We're so pleased to bring you another wonderful old title on Apiculture.

Not only is raw honey sweet and divine-tasting, it also has many health benefits including antiviral and antibacterial actions, wound-healing properties, and is packed with phytonutrients and antioxidants.

This special edition of **Productive Bee-Keeping : Modern Methods of Production and Marketing of Honey** was written by Frank C. Pellett, and first published in 1911 making it over a century old.

The book has sections on *Bee-Keeping a Fascinating Pursuit, The Business of Bee-Keeping, Making a Start with Bees, The Occupants of the Hive, Diseases and Enemies of Bees, Wintering,* and more.

A fantastic old text by an incredibly knowledgeable author, that makes this old book a great place to start for all Apiculture beginners, and anyone considering taking it up.

~ *Roger Chambers*
State of Jefferson, March 2018

To
THE MEMORY OF MY GRANDFATHER
B. F. CHAPMAN
FROM WHOM I RECEIVED MY
FIRST LESSONS IN APICULTURE

FOREWORD

THE author's earliest recollections are of days with his grandfather among the bees. One of the proudest days of his whole life was the first time he was permitted to cut a limb from an apple tree on which a swarm had clustered.

With a lifetime of intimate association with the bees and a wide acquaintance among the bee-keepers of the nation, it may not be regarded as surprising that he should undertake to set down in this book the information gleaned from so many sources.

In no other pursuit, perhaps, do the originators' names cling to the articles of equipment or methods of manipulation, as in bee-keeping. Most of the articles of equipment, as well as methods in common use, bear the name of the man with whom they originated—the Langstroth hive, Porter bee escape, Alexander feeder, Root smoker, Miller queen cage, and so on throughout the entire field of apiculture. So firmly established has this custom become, that a writer is in danger of being accused of plagiarism if he describes a method without the originator's name in connection. While the author has followed the usual custom, in the main, some methods have become so generally adopted that it hardly seems necessary to continue the practice. It is not with any intention of claiming as original any of these plans that the originator's name has occasionally been omitted, but rather because it does not seem needful with matters so fully credited already.

While the author believes that a few minor methods herein described are original with him, this book is not presented for the purpose of exploiting original material, but rather to de-

FOREWORD

scribe the accepted methods found valuable by extensive honey producers, under the greatest variety of conditions. The best has been gleaned from every possible source.

While most of the illustrations are from the author's original photographs or drawings made especially for this book, acknowledgment should be made for a number that are reproduced by permission from " Gleanings in Bee Culture," " The American Bee Journal," and other sources.

The author is also greatly indebted to Mr. C. P. Dadant, Dr. C. C. Miller, Dr. E. F. Phillips, and especially to Mrs. Pellett for valuable assistance.

FRANK C. PELLETT.

ATLANTIC, IOWA, November, 1915.

CONTENTS

CHAPTER	PAGE
I. Bee-Keeping a Fascinating Pursuit	1
II. The Business of Bee-Keeping	9
III. Making a Start With Bees	18
IV. Arrangement of the Apiary	36
V. Sources of Nectar	46
VI. The Occupants of the Hive	88
VII. Increase	100
VIII. Feeding	128
IX. Production of Comb Honey	136
X. Production of Extracted Honey	165
XI. Wax, a By-product of the Apiary	195
XII. Diseases and Enemies of Bees	206
XIII. Wintering	234
XIV. Marketing the Honey Crop	257
XV. Laws that Concern the Bee-Keeper	283

ILLUSTRATIONS

FIG.		PAGE
	The Orchard Furnishes an Ideal Location for the Apiary..*Frontispiece*	
1.	A Bee-Keeper Who Makes Pets of His Bees	2
2.	Getting Acquainted	3
3.	The Sting is an Effective Weapon of Defence	4
4.	Just for the Joy of It	7
5.	Many Successful Apiaries Built Up from a Single Colony	10
6.	A Few Colonies May be Kept on the Roof	11
7.	House Built from One Honey Crop from Less Than 300 Hives	12
8.	A Town-Lot Apiary	14
9.	Intensive Bee-Keeping	15
10.	The Silk Tulle Veil No Obstruction to the Vision	20
11.	A Youthful Beginner and the Necessary Outfit	21
12.	Good Hive Tools	22
13.	Smokers in Common Use	23
14.	Metal Top Covered with Flaxboard	24
15.	Tin Comb Bucket	25
16.	Observatory Hive	26
17.	An Apiary Ready for Shipment	31
18.	Transferring from Hollow Tree Without Cutting the Tree	34
19.	An Apiary Without Shade	37
20.	A Well-Arranged Apiary in California	38
21.	A Hive-Stand of Cement for Two Colonies	39
22.	A Tub of Water Covered with Chipped Cork Makes a Safe Watering Place	40
23.	A Long Trough with Burlap Lining for Watering the Bees	42
24.	The Bonney Hive-markers	44
25.	Soft Maple and Pussy Willow are Sources of Early Pollen and Nectar	53
26.	Catkins of Pussy Willow	54
27.	Blossoms of Soft Maple	55
28.	Fruit Blossoms Furnish Large Quantities of Honey for Early Brood Rearing	56
29.	The Golden Rod is an Important Source of Fall Nectar in Some Localities	62
30.	The Cup-Plant or Rosin Weed	64
31.	Blossoms of the Cup-Plant	65
32.	The Yellow Crownbeard is Much Sought by the Bees	66
33.	Wild Sunflowers are Important Honey Producing Plants over Large Areas	67

ILLUSTRATIONS

FIG.		PAGE
34.	Blossom, Seed Pod and Leaf of Partridge Pea	68
35.	Aster Honey Makes Poor Winter Stores	70
36.	Boneset or White Snakeroot	71
37.	Masses of White Snakeroot in the Author's Wild Garden	72
38.	Two Species of Heartsease or Smartweed	73
39.	The Horsemints are Valuable over a Large Scope of Country	75
40.	Catnip Yields Honey Abundantly	76
41.	Figwort or Simpson's Honey Plant	76
42.	The Rocky Mountain Bee Plant is a Valuable Honey Producer in Colorado	77
43.	Blossoms of the Button Bush	78
44.	Buckwheat in Bloom	79
45.	Where Sufficiently Abundant, the Wild Cucumber is Valuable	80
46.	Queen Laying in a Newly Made Comb	89
47.	Natural-Built Queen Cells	90
48.	Worker Bees at the Entrance of the Hive	92
49.	Drones	95
50.	Combs Showing Queen Cells and Capped Drone and Worker Brood	96
51.	Hiving Swarm in Straw Skep in Europe	103
52.	A Market Basket Swarm Catcher	105
53.	A Newly Hived Swarm	105
54.	Swarm Caught in a Sack, Running into the Hive	105
55.	Nuclei in Queen-Rearing Apiary	109
56.	Miller Queen Cage	114
57.	Benton Queen Cage	114
58.	Queen Cells by the Alley Plan	125
59.	The Minnesota Bottom Feeder	131
60.	The Miller Feeder	131
61.	Tin Pan Feeder in Super	132
62.	The Doolittle Division Board Feeder	133
63.	Metal Feeder After the Alexander Idea	133
64.	The Alexander Wood Feeder	134
65.	With This Entrance Feeder One Can See at a Glance How Much Feed Remains to be Taken	134
66.	Parts of a Comb Honey Hive	138
67.	Strong Colonies for Comb Honey Production	139
68.	Comb Honey Supers	140
69.	Comb Honey Super Dissected	140
70.	Sections for Comb Honey	141
71.	Separators for Bee-Way Sections	142
72.	Fence for Plain Sections	143

ILLUSTRATIONS

FIG.		PAGE
73.	Dr. L. D. Leonard Method of Putting Foundation into Split Sections	145
74.	The Pangburn Foundation Fastener and Sections Filled with Foundation	146
75.	Method of Putting in Foundation with Pangburn Fastener	147
76.	The Use of Super Springs	148
77.	Ventilated Bee Escape and Queen Excluders	153
78.	The Porter Bee Escape	161
79.	Sphuler's Hand Extractor as Used in Europe	166
80.	Storage Tanks of a Large Honey Producer in California	167
81.	A Power Driven Extractor	168
82.	Sixty Pound Cans for Extracted Honey	170
83.	The Townsend Uncapping Box	171
84.	The Peterson Capping Melter	172
85.	Bingham Uncapping Knife	173
86.	Langstroth Hive for Extracted Honey	174
87.	Langstroth Hive Dissected	174
88.	A Well-Arranged, Two-Story Honey House	176
89.	Large Honey House With All Work on Ground Floor	177
90.	The Automobile is Valuable for Outyard Work	178
91.	Upper Comb Built on Full Sheet of Foundation; Lower Without Foundation	180
92.	Usual Method of Wiring Frames	181
93.	Hoffman Frame with Full Sheet of Foundation	182
94.	Development of Combs from Foundation	182
95.	Comb Built on Wired Frame with Full Sheet of Foundation	183
96.	Strong Colony for Extracted Honey Production	184
97.	Colony that Produced Forty Dollars Worth of Extracted Honey in One Season	184
98.	Wheelbarrow Load of Extracting Supers	188
99.	Utilizing Feed Cooker for Liquefying Candied Honey by Steam	192
100.	Hatch Wax Press	202
101.	Steam Wax Press	204
102.	Brood Comb from Colony Affected with American Foul Brood	208
103.	Work of Wax Moths in Colony Affected by American Foul Brood	209
104.	Thirteen Colonies Left of One Hundred Five as the Result of European Foul Brood for Eight Months	216
105.	Appearance of Larvæ Affected by European Foul Brood	218
106.	The Natural and Preferred Food of the Skunk is Insects. The Honey-Bee is a Tempting Delicacy to the Skunk Palate	225
107.	The Robber Fly	226
108.	The Value of a Good Natural Windbreak Behind an Apiary Can Hardly be Overestimated	238

ILLUSTRATIONS

FIG.		PAGE
109.	Paper Winter Cases Are at Best Scant Protection, But Are Good for Cellar-Wintered Bees After They Are Placed on the Summer Stands	241
110.	The Dadant Method of Outdoor Wintering in Large Hives is Suited to Localities Where the Bees Have Frequent Flight During The Cold Months	242
111.	One Method of Packing on the Summer Stands	243
112.	Parts of a Double-Walled Hive	244
113.	Double-Walled Hive Assembled	245
114.	Packing Box with Hives Inside Ready for Leaves or Other Packing Material for Outdoor Wintering	248
115.	Packing Two Colonies with Dry Leaves in a Goods Box	249
116.	Snug for the Winter	250
117.	The Packing Boxes May be Utilized for Chicken Coops in Summer	252
118.	Concrete Cellar for Wintering	253
119.	Cellar for Wintering Under the Workshop	254
120.	Development of Comb Honey in Sections	259
121.	Packages for Retailing Extracted Honey	264
122.	Trade-Mark of the Colorado Honey Producers Association	265
123-126.	Honey Labels	266, 267, 268
127.	Little Stickers Widely Used for General Advertising	269
128.	Advertising Sign at the Bonney Apiary	270
129.	Iowa Bee-Keepers' Association Holiday Placard	271
130.	An Exhibit at the Fair is a Good Advertising Medium and Promotes the Use of Honey	272
131.	Paper Carton the Best Retail Package for Section Honey	273
132.	The Hunten Tin Package	273
133.	Dr. Bonney's Postcard Which Brings Him Many New Customers	279
134.	The Automobile as a Sales Agency is the Most Up-to-Date Method	281

PRODUCTIVE BEE-KEEPING

CHAPTER I

BEE-KEEPING A FASCINATING PURSUIT

While this book is written for the purpose of encouraging honey production as a business enterprise, and, accordingly, deals with the subject in a very practical manner, the reader is asked to allow the suggestion here at the beginning, that there is much of poetry, as well as hard work in making a living from the apiary.

Honey-Bees as Pets.—No, this is not a joke, for bees really do make nice pets. They are always interesting, and have this advantage over most other pets: they can be left to look out for themselves without inconvenience during their owner's absence. While there are comparatively few who keep bees as a sole source of livelihood, there are many thousands who keep a few colonies for a diversion, as a side line, or for the fun of the thing. Yes, it is safe to say that nearly every really successful bee-keeper comes to feel a strong affection for the busy little insects, and to regard his bees as pets (Fig. 1).

To nature lovers, the pleasure of association with the bees outweighs the pleasures to be bought with the cash realized from the sale of the honey; hence they cannot refrain from growing very enthusiastic about bee-keeping as a business, and sometimes the enthusiasts are accused of painting the picture with too much bright color. Perhaps some such feeling is essential to the pursuit, and the lack of it may account for the failure of some, who are not lacking in industry or patience, two very essential requirements.

Getting Acquainted.—If one will make pets of the bees, he must first proceed to get acquainted with them. They are notional little creatures, and one must know what to expect under given circumstances in order to get along well. One who loves and understands bees seldom has trouble on account of stings. The sting is a weapon of defence, seldom of offence, and the bee-keeper must know the liberties they will resent (Fig. 2).

Fig. 1.—A bee-keeper who makes pets of his bees.

Of course there is a difference in the disposition. Some bees are crosser than others, and, perhaps, there are bees which one would hardly care to cultivate as pets. The author has at different times had a great many colonies of Italians, Crosses, and Blacks. Some have been gentler than others, but he has usually been on friendly terms with all. The practical bee-keeper will frequently handle his bees without veil or gloves, and without a sting. Others, who have had a few bees about for years, without really becoming acquainted with them, always arm themselves with a sting-proof armament, and usually arouse the bees to such an extent that it is unsafe for any member of the family to leave the house for twenty-four hours. At such times

FIG. 2.—Getting acquainted.

chickens have been known to be stung to death, and other animals to be badly used.

The successful bee-keeper must take the trouble to get acquainted with the bees, and to comply with the few simple requirements necessary to handle them easily and successfully. In the first place, never place yourself in the direct line of flight of the workers, in going to and from the hive. People who should know better are often seen getting directly in front of the hive, even though a rod or two away, to watch their movements. An expert called upon to look into a hive may, by approaching from the rear, carefully remove the cover without causing any commotion. At the same time the novice, watching from some distance in front, is quite likely to receive sufficient attention to insure a hasty retreat (Fig. 3).

FIG. 3.—The sting is an effective weapon of defence.

An Orderly Community.—The work of the hive is done in an orderly manner. There is no hit-and-miss business there. Every individual bee has a duty to perform, and that duty is apparently done in the right manner and at the proper time. In order to look within the hive without causing resentment on the part of the bees, one must do something to break up the orderly system and create confusion among the inmates. Under normal conditions, sentinels are posted at the entrance of the hive to detect and ward off danger. In some manner these guards are able to recognize every member of the very numerous family. If a strange bee, a robber perchance, should happen to alight at the entrance of the hive, it is at once set upon and driven away or killed. Let a man or an animal pass in front of the hive, and the chances are that the sentinels will take notice, and invite the trespasser to move on. The bee-keeper, wishing to open the hive, approaches quietly from the rear, and blows a little smoke into the entrance. As a result the sentinels are at once thrown off guard. The cover is then carefully lifted and more smoke blown over the frames. This causes a suspension of work in all parts of the hive, and general confusion results. The bees at once seek the open cells, and fill their honey sacs with honey, as though they believed the house to be on fire and wished to save as much of their hard-earned store as possible.

A careful operator will be able to create such a condition of hopeless confusion within the hive, that the bees lose all thought of defence, and he can handle them at will without the slightest resistance. If the frames are at once removed, the bees may be dumped into a pan, picked up by handfuls, or disposed of in any manner, if only one be careful not to pinch or crush any of them. Experienced bee-keepers frequently give demonstrations before the gaping public in a manner to excite a wondering interest on the part of the uninitiated, and to lead to all sorts of absurd statements. Some go so far as to attempt to give the impression that they have unusual influence over the insects, calling themselves bee-wizards or other silly names. If the operator is skillful in controlling the bees, he can perform feats that seem very wonderful to those whose only information concerning them is that they sting and make honey. Blowing live bees from the mouth, pouring panfuls over the head, and similar "stunts" are not uncommon at these demonstrations.

There are some gentle strains of Italians that have become so accustomed to being handled that they can be safely handled during a honey flow without smoke. The novice should be cautious about over-confidence until he has become familiar with the habits of the insects and the methods of control.

There are some who cannot overcome a nervous fear of the bees, and consequently can never handle them successfully. The first essential in controlling bees is to be able to control one's self. When a bee comes buzzing about, the chances are ninety-nine in a hundred that she will make no trouble unless the person under observation starts it. How often people get stung by starting a fuss with a perfectly friendly bee, when if they would only keep quiet there would be no trouble. One can very soon come to recognize the difference between the hum of a friendly bee and the angry buzz of one on the warpath. The experiment has been tried of keeping perfectly still when pursued by angry ones. Often they alight on the operator with apparent surprise that he is not kicking up a fuss, and, after a moment or two of hesitation, fly away without drawing their daggers. This plan is not always successful, though there is less danger of getting stung when quiet than when frantically kicking and striking in every direction. Where a colony is on the warpath, the best plan is to keep away until they have become quiet, for it is very difficult to control bees after they have become fully aroused.

Fifty or a hundred friendly bees crawling over a seasoned bee-keeper cause him not the slightest uneasiness, but on the other hand, he is likely rather to enjoy the sensation. One who is not accustomed to handling them should always take the precaution to protect himself fully with veil and gloves, until he becomes so familiar with them as to be able to overcome his nervousness when they alight on the face or hands.

Some Causes of Trouble.—There are several things that have a tendency to cause trouble between the operator and his bees. They are much more inclined to be cross when the atmosphere is heavy before a storm, and sometimes after. They show a tendency to be more hostile toward one dressed in dark colored clothing than in light garments. One should take care to never go about the bees with the odor of the stable clinging to his garments, as that is offensive to them. One is more likely to be stung when perspiring freely, and persons whose perspiration has an offensive odor will have more trouble with the bees.

One who is much with the bees can, if he will, soon come to know and avoid the things that are distasteful to them, and to perform the operations necessary to bee-keeping with little danger of being stung.

The Joy of It.—The nature-lover who does not keep bees is missing a good thing. There is a charm about lying in the grass

FIG. 4.—Just for the joy of it.

beside the hive and watching the stream of workers bringing in the harvest of honey and pollen at the height of the season, when the colony is in a fever of excitement. Then to know something of the wonderful system of government, by which the thousands of insects composing a colony are able to work together harmoniously, with never a shirker among the bevy of toilers, is a most interesting study. At times the bee-keeper is seized with a desire to see what is going on inside the hive, to visit a colony, remove the frames, and examine the young bees in all stages of development, hunt out the queen, pick up handfuls of the friendly little bees just to feel the tickle of their feet in his hand, and to put them all back again, just for the joy of it (Fig. 4). Yes, indeed, it is worth while to make pets of the bees.

QUESTIONS

1. Note some of the attractions of bee-keeping.
2. What are some of the essentials of success?
3. Discuss the general principles of bee control.
4. What are some of the things that are distasteful to the bees?

CHAPTER II

THE BUSINESS OF BEE-KEEPING

Few persons think of bee-keeping as a business. The ordinary conception is that of a diversion, a side line on the farm, or a harmless pursuit for old men. Perhaps 90 per cent of those keeping bees may be included in one of these classes, of which a very large number will come under the head of keeping bees as a diversion.

The public is just now beginning to realize the fact that bee-keeping is a real man's-sized job, and that an able-bodied man of good education can profitably occupy his time with bees.

When considering the possibilities of any occupation as a lifetime pursuit, the careful person makes inquiry along several lines: Is the business congenial? What are the advantages? What are the probable returns?

No specialized branch of agriculture requires more skill to be successfully pursued as an exclusive business than honey production. The man who cannot or will not give close attention to details, promptly, should never be a bee-keeper. The whole business is one of details, and apparently unimportant things are of the utmost importance. To such an extent is this true, that it often happens that the scientific bee man will get a crop of honey in an off season, when his neighbor, with the same kind of equipment and apparently following the same general plan, gets no surplus. In most localities the honey flows are of short duration, and everything hinges on getting the bees in proper condition to store the maximum of honey when the flow is on. The honey producer must see to it that his dish is right side up when it rains nectar.

The man or woman who is of a studious disposition, loves nature, and delights in out-of-door pursuits, is likely to find bee-keeping a congenial occupation. Most of the conspicuously successful bee-keepers are studious, questioning individuals, intensely interested in the honey-bee. While great progress has been made in the past few years, much yet remains to be learned, and new methods and new discoveries are constantly brought forward. The person who believes he knows all about bees is a back number, indeed.

The Advantages.—Bee-keeping is one of the few pursuits open to persons of small capital or poor health. Many a success-

Fig. 5.—Many a successful apiary has been built up from a single colony.

ful apiary has been built up from a single colony of bees and an investment of but a few dollars (Fig. 5). In fact, some of the most successful bee men have begun in this way, and built up an extensive business that yielded a good income.

Then again, bees may be kept in situations where it would be impossible to undertake any other enterprise. Of course, after one has enlarged his apiaries to such an extent that they will occupy the entire time and attention of the owner, a suitable situation will be necessary, but a start may be made under apparently unfavorable circumstances. A few colonies are often kept on top of a business building in the city, in the attic, the back yard, or even have been known in the bed-room, with an opening through the sash (Fig. 6).

One of the greatest advantages of the business lies in this possibility of development, without requiring that the learner

Fig. 6.—A few colonies may be kept on the roof.

leave his regular home or business until he has learned much concerning the new venture and is able to judge whether he is likely to be adapted to the work. Men and women, worn out with professional work, and feeling the need of change and of work in the open air, have found health, happiness, and prosperity in following this suggestion (Fig. 7).

Women in many cases are successful honey producers, those who have laid aside the arduous work of the school-room to take it up being not uncommon.

12   THE BUSINESS OF BEE-KEEPING

The fact, perhaps, that so many in poor health or otherwise unfortunate have taken to bee-keeping may be in part responsible for the general impression that, as a business, it amounts to little. The writer knows many men of perfect health, good business ability, and other qualities that contribute to success in any calling, who are devoting their time and energies to this business,

Fig. 7.—House built from one honey crop from less than 300 hives.

and it is from the inspiration of their success that he hopes to draw for whatever of merit this book may possess.

The Returns.—A most important consideration is the financial return, for expenses must be met, families are to be supported, and most of us must have a care to make ends meet. While there are those who keep bees in a very large way, with a series of many outyards and much help, it is rather the one-man business that we will just now discuss, for many people who can be successful in a business whose every detail they can oversee are likely to fail when it comes to organizing a system and delegating the actual operations to hired help.

A Minister.—As a first example there is the case of a Presbyterian minister, who took up bee-keeping exclusively several years ago. He now has about three hundred colonies of bees, in four yards. One hundred and twenty colonies is the largest number that he has in one yard, while there are but thirty colonies in his smallest yard. His average return has been seven dollars per colony per year. This amount is somewhat in excess of the salary he probably received in serving a small congregation in a country town.

A carpenter gave up his trade to keep bees as an exclusive source of livelihood, more than twenty-five years ago. When he abandoned his trade and took up bee-keeping he rented a house and two lots in a small town. At the end of two years he purchased the property, and has since occupied it as a home. During the early years of his experience before he became well established, there was one season of failure of the honey crop, when he found it necessary to work at his trade temporarily for a few months. Aside from that, the bees have furnished his entire support. He has paid for his home and business, from the apiary, built a better house, and added to his real estate holding. While his income is not large, he has had a better support than his trade could furnish, and his business is at home where he enjoys the assistance and association of his family. His work is of a kind that he enjoys, and not of a nature that advancing age will compel him to lay aside (Fig. 8).

A Clerk.—One might also cite the case of a shipping clerk in a manufacturing establishment. Because of failing health he was compelled to seek the open air. The pressing necessity of providing for his family compelled him to find something that would furnish the needed support, without demanding too heavy toil from a weakened body. He has been remarkably successful

Fig. 8.—A town-lot apiary that has been its owner's sole dependence for more than twenty-five years.

considering his circumstances, and now feels that the condition that compelled him to make a change has proved a blessing in disguise.

A book-keeper in a western city has for some time been

Fig. 9.—Intensive bee-keeping. Corner of an apiary where 165 colonies are kept on lot 60 x 110.

developing his business to the point where it will justify him in cutting loose from his salary and devoting all his time to honey production. He has grown up in the work so gradually that he has reached the point where he can make the change without feeling the cost, as the bees paid their own way, and without feeling the shock of readjustment. He lives out on a car-line, where he has two or three lots. He has been attending to his regular duties at the desk, and giving his evenings and mornings and occasional holidays to the bees, assisted by an enthusiastic wife (Fig. 9). One season he produced and sold more than fourteen hundred dollars' worth of honey, which quite probably was equal to his salary. Should he decide to devote all his time to the bees, he can care for double his present number. While this was an unusually favorable season, with double the number of colonies, his average production will leave little risk to run.

A General Farmer.—One of the most successful bee-keepers of the Middle West is a young man who abandoned general farming because the heavy expenses necessary to pay cash rent, hired help, buy expensive machinery, and replace the worn-out horses made it difficult to get ahead. This man does nearly all his own work, thus keeping down expenses. He produces from twenty-five thousand to forty thousand pounds of honey per year, which he sells to jobbers at wholesale prices. By developing a retail market he could increase his income materially, though it is good at present.

Many Others.—It would be possible to multipy these examples indefinitely, but these men who have turned to bee-keeping from so many different walks of life should be sufficient. It would be possible to cite also numberless examples of those, who, by plunging without experience, have failed, but most of the failures have been because the adventurer did not use good business judgment.

As an Exclusive Business.—The men who are engaged in honey production as an exclusive business are getting results equal to those derived from other lines of agriculture, with less capital invested and with less risk. The fact that the business is open to men of small capital, who are unable to engage in general farming because of the larger outlay required, surely makes it desirable to encourage the development of the industry as far as possible. Bee-keeping, as a business, requires high-grade talent, and comparatively few men succeed in making it profitable as an exclusive line. This is not the fault of the business but of the men. It looks so easy that men are not willing to serve an apprenticeship, or to take the necessary time to master the business in all its details, as they would expect to do in other lines.

Judging from the incomes of those who are depending upon bee-keeping for a livelihood, it seems safe to say that a man who will become thoroughly proficient and attend properly to his business can make from twelve hundred to three thousand dollars per year from the bees that he can care for personally. Some do better than that, many do not do as well, but so many exclusive bee-keepers come within this range that it is a conservative one. If the ambitious reader proves to be the exceptional man, he may hope to increase his income much beyond the higher figure by skilful organization and large apiaries widely scattered.

After gathering the average results from a number of bee-keepers who have kept bees for many years, it seems safe to place the average return in the average locality at five dollars per colony in the hands of expert bee-keepers. So much depends upon a suitable locality that it is important that one who is taking up bee-keeping as a business should choose a locality above the average if possible.

The Outlook.—There are always a few timid souls who cry over-production, who feel that the honey business will shortly be overdone. The last census clearly shows that there are a less number of bee-keepers in the United States than there were ten years ago, although there has not been a corresponding decrease in the number of bees. This indicates that the bee-keepers are becoming specialists. When it is remembered that there has been a constant increase in population, one need have little fear of over-production of honey while the number of bee-keepers is decreasing, especially not until we reach the time when there is a marked increase in the production of honey. While at times there may be a temporary glut in some markets because of improper distribution, the bee-keeper in taking up the business need have little fear of seeing the production of honey overdone for many years to come.

QUESTIONS

1. What type of person is most likely to be a successful bee-keeper?
2. Note some of the advantages of bee-keeping as a business.
3. Compare the returns of bee-keeping to other occupations requiring equal capital.
4. What is the outlook for the business?

CHAPTER III

MAKING A START WITH BEES

Unless one has had rather extended experience and observation, it is nearly always advisable to begin with only one or two colonies and grow into a business as extensive as inclination or opportunity will permit.

Proper Equipment.—Only a small percentage of bee-keepers start right and select equipment that will continue satisfactory. Hundreds of men have started with hives or other equipment unsuited to their locality or the system that they have chosen to follow, which later caused a heavy expense to change. Not long since the author visited a young man who is employed in a large machine shop. His spare time is taken up with his bees, to which he hopes before long to give his entire attention. He has been very fortunate in making his selection of equipment, for everything which he has purchased is likely to prove of permanent value. His hives are of the best, his combs are straight and built on wired frames, and everything indicates the bee-keeper of long experience, instead of a beginner.

Getting Experience.—If one is so situated that he can do so, it is very desirable to spend at least one season in a large apiary. This is not only very desirable to any one who expects to make honey production a business, but doubly so to one who wishes to start on a liberal scale and increase rapidly. One should select the most successful bee-keeper, of whom he can learn under similar conditions to which he expects to work. Systems that are adapted to one locality may fail in another. To serve such an apprenticeship is not altogether essential, for many successful bee-keepers have developed their own systems from their own experience, with the help of ideas gleaned from the bee journals and books relating to the subject. A course in bee-culture in one of the agricultural colleges offering such a course is very desirable.

A Beginner's Equipment.—Hives of the Langstroth dimensions are now almost universally recommended, because of the fact that they are everywhere standard. Hives of other patterns may be equally good for practical service, but the purchase of supplies may be difficult, bees offered for sale in them may bring much less because of the fact that the buyer will want them in standard hives, and similar reasons. Supplies for the standard hives can be secured almost anywhere, and bees in such hives are usually saleable in localities where bees can be sold at all.

On the other hand, there is a decided difference of opinion as to the size of hive. In many cases the eight-frame hive has been selected, only to prove too small. This small hive body, which is largely occupied for brood-rearing, is too small to accommodate a vigorous queen, and forces much of the honey into the supers during the honey flow, with the result that in many localities, where the flows are short and rapid, insufficient honey remains in the broad chamber for wintering. In the hands of inexperienced persons many bees are lost from lack of stores. The necessity of feeding at the close of the honey flow requires a lot of work and is not always agreeable, as the author has found by experience. While many persons have changed from the eight-frame to the ten-frame size, but few successful bee-keepers have changed from the ten-frame to the smaller size. In some localities, hives of this pattern as large as twelve-frame are in use. Most of the successful men prefer the ten-frame, and it would seem to meet the requirements of a greater number than any other size. In few localities does the eight-frame hive seem to be suited to conditions. Seldom does one find an experienced man working with hives of other patterns but who is free to say that they have been a source of annoyance, to say the least. Of course if one is situated where some other hive is in almost universal use, the advantage of having equipment similar to that in general use would be an item not to be overlooked. The Dadant hive has some advantages over the Langstroth hive, especially for extracted honey production. This is the standard hive in parts of Europe, but its use in this country is restricted to a few localities. The tendency of the times is more and more toward the large hive. Possibly from the one extreme, popular favor may go to the other, and it seems wise to caution against either the extremely large or too small hive. The ten-frame Langstroth would seem to offer a safe intermediate.

Tools for the Apiary.—The beginner, even though he have but one hive, will need a good veil and gloves, a suitable hive tool, and a smoker. Cotton flannel gloves with long gauntlets are

Fig. 10.—The silk tulle veil offers no obstruction to the vision.

most satisfactory for use in the apiary. Rubber or other heavy material will be disappointing in results, as well as much more expensive.

Veil.—A good veil is one of the most necessary articles of equipment (Fig. 10). One who is not a seasoned bee-keeper should not risk going much about the bees without perfect protection. When one has come to understand the peculiar habits of the insects, he will know when it is safe to work without protection, and when he should stay away, but the beginner is very likely to be severely punished most unexpectedly. There are many different kinds of veils in the market. As a rule the most expensive give the least satisfaction. The globe veil, which is listed in nearly every catalogue, is a nuisance and seldom used by extensive honey producers. A satisfactory protection can easily be provided by sewing mosquito netting to the rim of a straw hat. This, however, catches on every twig and is easily

Fig. 11.—A youthful beginner and the necessary outfit.

torn. The Alexander veil is one of the best, though the one shown in Fig. 11 is as good as any. This is made of a strip of screen wire rolled into a cube. A cloth is sewed over the top, and an apron about the bottom, which is easily tucked under the coat or suspenders.

Hive Tool.—While the man with a few colonies can get along with a screw-driver or chisel, a suitable hive tool is very convenient and helpful. To the large honey producer it is essential, for the saving in time will pay for it within a few hours, in the busy season. There are several styles on the market, each with its peculiar advantages. The hive tool should be so constructed as to serve as a pry in loosening frames, have a sharp surface to scrape off burr combs, propolis, etc., and at the same time be small enough to handle easily and quickly (Fig. 12).

A smoker is essential and should be procured with the first colony of bees. A little smoke, intelligently applied, will enable one to control the bees so nicely that it is very unwise to do with-

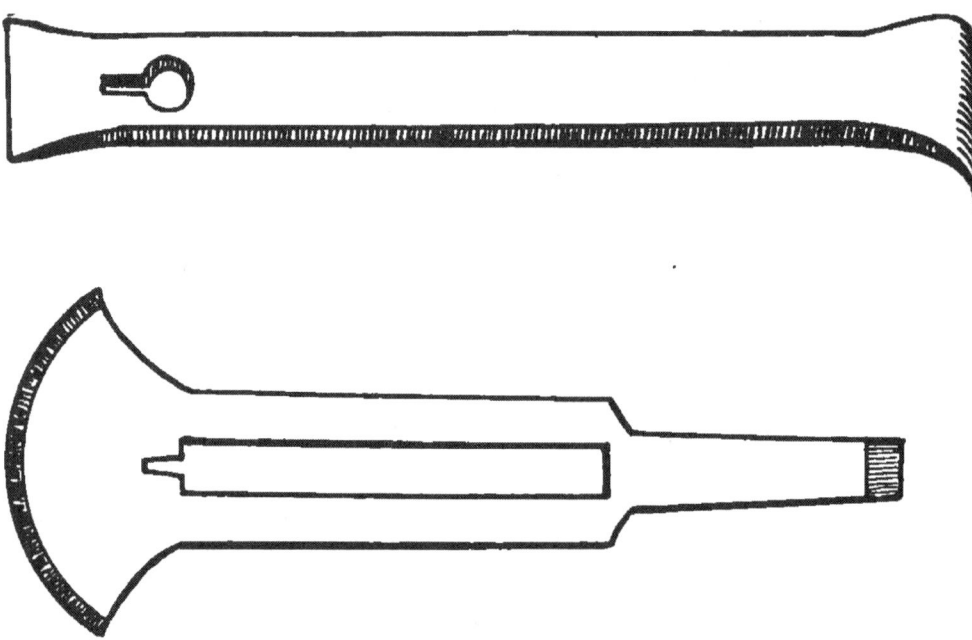

Fig. 12.—Good hive tools.

out it for a day. There are two very excellent kinds on the market and several indifferent ones. Most beginners make the mistake of buying a small size, because they have only a few colonies of bees. The larger size costs but a few cents more, and is much to be preferred in every way (Fig. 13).

Rotten wood is a very satisfactory smoker fuel, although excelsior, cotton rags, greasy waste, or any similar material will do.

Care should be taken not to use too much smoke, a very

common fault with beginners. If one has gentle bees, a very slight puff at the entrance and then another over the frames when the cover is removed will be sufficient. If the bees are inclined to be cross, a little more may be necessary. The tendency is rather to use too much than too little. The use of smoke is very disturbing to the bees, and the successful apiarist interferes with the normal condition of the colony as little as possible. Every disturbance during the honey flow must be accounted for in honey stored.

Fig. 13.—Smokers in common use

Minor Equipment.—There are many things for use about the apiary which, while very necessary in themselves, make no difference in results as to which particular kind is adopted. In these minor items there is room for unlimited argument to no profit. The particular article that best suits the individual taste is the one to adopt.

Covers.—A good cover is very essential, but which is the best will depend a great deal on who is deciding. A flat wood cover with a strip at each end to prevent warping is very satisfactory. The piece covers are made of such light material that they do not, as a rule, last as long as is desirable.

More and more are the metal top covers coming into general favor. These covers are flat topped and made of a sheet of galvanized steel or iron on a wood frame that telescopes over the top of the hive. A thin inner cover is used under them. This makes an air space of nearly one-fourth inch between the inner cover and the corrugated paper or board, with which the cover is lined. The telescope feature makes the cover much less likely to be blown off during high winds.

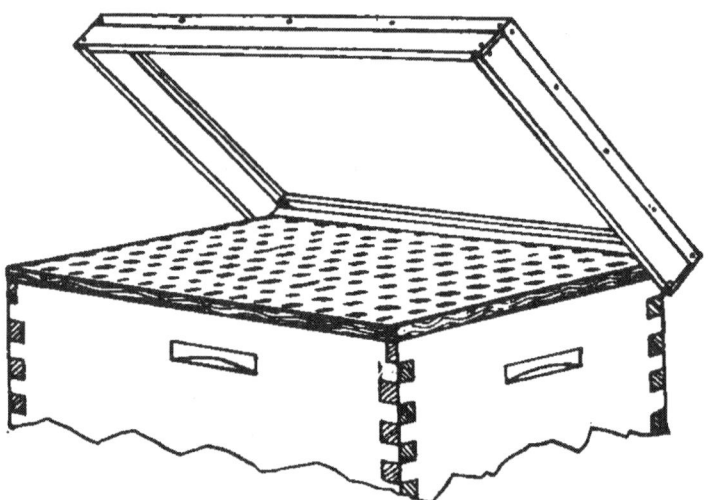

Fig. 14.—Metal top cover with flaxboard.

The chief objection to these covers is the fact that they get very hot when the hive is in direct sunlight in warm weather. If sufficient provision for ventilation is made, the effect will not be so noticeable. If painted with a light-colored paint, less heat will be absorbed than if painted some dark color. Dark-colored paints are not suitable for bee hives. Fig. 14 shows a metal top cover with flaxboard used above the inner cover.

Flaxboard is a new absorbent and insulating material which has recently appeared in the market. It is composed of flax fiber pressed into sheets of suitable thickness. Although not yet extensively tested, it seems to give excellent satisfaction where tried. A sheet of this material one-half to three-fourths inch in thickness used as a lining for metal top covers not only prevents the hive from becoming overheated in summer but absorbs surplus moisture in winter, and also retains the heat of the cluster. It bids fair to come into general use. Flaxboard should only be used under a water-tight cover, as it will quickly absorb any drip, and, if the cover leaks, will soon be ruined.

Comb Bucket.—A comb bucket is a very useful article in even a small apiary (Fig. 15). It is a convenient way to carry a few combs when making nuclei or equalizing brood. A tight-fitting cover is an advantage when there is a tendency for robbers to be prying into every opening. The one shown in the illustration is of tin, but some bee-keepers make them at home, of thin boards, which not being subject to rust will be more lasting.

Observatory Hive.—Every bee-keeper should have an observatory hive for the purpose of becoming familiar with the habits of the bees (Fig. 16). These hives are made with glass sides, so that one can see what is going on inside. Some are made full size, but as it will then be impossible to see the interior of the brood nest, they are not very satisfactory. The most suitable is the single frame observer, which can easily be made by anyone handy with tools, or can be bought of any dealer in supplies.

After the weather has become warm, a single frame of brood and bees can be taken from any hive where it is desirable to replace the old queen. She is taken with this frame and placed in the observing hive, where the curious bee-keeper can see every move, and spend many idle hours profitably in watching his pets. The queen is thus easily observed while laying, the young bees can be seen during all stages of their development, and the field bees can be seen bringing in their loads of pollen and nectar and depositing them in the proper place.

This little colony will carry on all the usual activities in a normal manner under the very eye of the bee-keeper. If desired the little hive can be placed in the living room with an opening through the sash, for the bees to go to and from the fields.

Fig. 15.—Tin comb bucket.
Fig. 16.—Observatory hive.

Buying Bees.—It is usually best to buy the bees within easy reach of the place where they are to remain if possible. The expense of shipping long distances with the consequent danger of mishap and loss are thus eliminated.

As a rule, unless one is willing to pay a fancy price, he need not pay much attention to the kind of bees, providing the colony is a strong one. The best way to improve the stock is to buy a queen from some reliable breeder, and after killing the old queen and leaving the colony queenless two or three days, introduce the new one, following the directions that accompany her.

In buying one should pay according to the condition of the colony. If the bees be common stock, in box hives, the price should not be high, as it will be necessary to add the further expense of a suitable hive and the labor of transferring, which is never an agreeable task. If the bees be in a good hive, on straight combs in good condition, the price may then be much higher, for they are ready for business when the honey flow begins. It too often happens that bees for sale in good hives have received no attention, with the result that the combs are built crosswise, making it impossible for the operator to get into the brood nest without disastrous results. Such colonies will also have to be transferred, which will add considerable to the cost.

In order to conduct bee-keeping profitably, it is necessary to have every comb in every hive so that it can be easily removed for the purpose of examination or exchange. It frequently happens that for one reason or another the bee-keeper must take combs of honey or brood from one colony to add to another, or he must examine the interior to ascertain the condition of the colony. Successful honey production is absolutely impossible unless conditions are such that the bee-keeper can reach the farthermost corner of the hive when necessary.

A colony of pure Italian bees, on straight combs, wired frames, in good ten-frame hives without too much drone comb, is cheaper at eight or ten dollars than a common colony in a box hive at a dollar. Especially is this true in the spring of the year.

when the one is ready for the honey flow, while the other must be transferred and much of the season lost in building up to the point of storing surplus.

A set of ten good brood combs in wired frames is worth at least two dollars. A new ten-frame hive, complete, will cost three dollars or more, and an Italian queen another dollar. This does not leave a great deal for one's labor in transferring, so that the colony ready for business is likely to store more than enough additional honey to make up the difference.

However, in buying bees, unless one is prepared to ship for a considerable distance, he sometimes finds it necessary to take what happens to be offered.

Moving Bees.—If one will go to the apiary on a warm day when the young bees are taking their first flight, he will observe with what care they mark the location of the hive. At first they fly but a few inches from the entrance and pass back and forth many times, always facing the hive. Each time they gradually lengthen the line of flight, back and forth, up and down, until they have received an indelible impression of the appearance of their home. After they have fully examined the front of the hive they fly a little farther, until they can get a similar view of the immediate surroundings at a distance of a few feet. The flight now takes the form of irregular circles, which are gradually enlarged to take in the apiary and in time the whole country roundabout. These preliminary flights are always taken by the young bees, before they take up their duties as foragers in the field.

Apparently they come to depend entirely upon the sense of location thus developed, and afterwards fly directly to the hive entrance from any point of the compass, with little attention to anything but the location. If the hive is taken away and another set in its place, they will enter the new hive without a moment's hesitation. Once inside they discover their mistake, and hurriedly tumble out and take to flight. After a moment's examination they reassure themselves that the location is correct, and re-enter the hive again.

If the hive is moved but a few feet away, they are greatly confused and will require some little time to accustom themselves to the new location. If the day be warm and the workers are in the field, hundreds of them will soon be flying about the former location of the hive.

Bee-keepers take advantage of this characteristic of the honey-bee to return to its old location regardless of changes, to make swarms hive themselves. The queens are clipped so as to be unable to accompany the swarm, and the bees, missing her, return to the old home only to find it gone and an empty hive in its place, as described in Chapter VII.

If bees are only moved a short distance many will be lost by returning to the old stand, unless some precaution be taken to insure that the new location will be carefully marked by all bees leaving the hive. For this reason it is best to move the bees three or four miles if possible. When they are moved a less distance it is well to place the hive in a dark cellar for several days; a week if they can be kept quiet that long. After the hive is taken from the cellar and set in the new situation, it is well to turn a large box over it, and remove a board near the ground to make a decided difference in appearances to the bees coming from the hive. They will then be likely to take note of the new location, and return in safety to the hive. After a day or two the box can be removed. Its only object is to create a new appearance. Bees moved for a considerable distance find conditions so strange, that there is little danger of loss from failure to return to the hive. The shorter the distance, the greater the difficulty in moving them, unless it be when they are removed from the cellar in spring, when they can safely be placed in any situation. However, even then, if they are only placed a short distance from the old stand, some of the old bees will return to the place where the hive stood the fall before. If the bees are to be taken but a short distance, say two or three rods, it is a common plan to move them a foot or two each day. They quickly adjust themselves to such a short move. While this plan is tedious, nevertheless it is safe.

Ventilation.—In hot weather great care must be taken to see that the bees have sufficient ventilation when confined to the hive. In cold weather, a wire screen over the entrance will be sufficient, if the bees are to be moved but a short distance. In summer the cover must be removed and the top covered with screen also (Fig. 17). Sometimes even this is not sufficient and strong colonies are likely to be lost. It is well to avoid moving bees in very hot weather if possible. When it becomes necessary to screen the top of the hive, a frame should be used that will raise the screen a little above the top, thus providing an empty space above the frames. If they show a tendency to crowd about the entrance and against the screen on top, a little water sprinkled over the surface will serve to quiet them.

To Tell Strong Colonies.—In buying bees in late fall when a long winter is ahead, colonies heavy with honey as well as strong in bees should be selected. After making allowance for weight of hive, bees, etc., there should be at least twenty-five pounds of honey in the hive, and forty is better to insure an abundance of stores for spring brood rearing. Some bee-keepers figure that fifty pounds is not too much to leave in the hives for winter.

If one buys bees in spring, which is the best time for one making a start in bee-keeping, it is well to select them during the period of fruit bloom. In the ordinary apiary at this season of the year, colonies will vary greatly in condition. Some will be very strong and some very weak. Then there are likely to be queenless colonies, which one would not care to buy at any price. The strong colonies are the ones to look for, for the weaklings are likely to be so slow in building up that they will be of little value in storing surplus, unless they receive special attention.

By walking through the apiary on a warm day at this season, when the bees are active, one can readily pick out the strongest colonies from the appearance at the entrance of the hive. The colonies showing the greatest activity at the entrance, especially if the workers are carrying in large quantities of pollen, are the ones to mark for further examination. The pollen balls are very conspicuous on the legs of the workers, and one can thus see at a glance something of the condition of the colony. The pollen is mixed with honey, and used to feed the young bees. At this season large quantities of brood insures a strong working force

Fig. 17.—An apiary ready for shipment.

a little later to gather the principal honey crop. As a rule, the colonies bringing in the most pollen will be found to have the most brood.

After deciding from external appearances which colonies are worthy of further examination, the hives should be opened to see that the combs are straight, that there is a sufficient quantity of honey to last until the main honey flow commences, and that not too much drone comb is present, as this will necessitate its removal to be replaced with full sheets of foundation. There should, ordinarily, be about ten or fifteen pounds of honey in the hive at this season, to insure the safety of the colony. It sometimes happens that a rainy season during the first days of the main honey flow will result in the starvation of strong colonies, with the hives full of brood. Five or six frames of brood consume a surprising quantity of food, and a short period of time during which nothing is coming from the field causes the bees to draw heavily on their stores. If no stores are present the result will be disastrous for the bees and for the owner as well.

Transferring.—The old books on bee-keeping usually recommend the cutting out of the combs containing brood and honey and fitting them into the frames, tying them in with cotton strings. The bees will shortly fasten the combs and remove the string. While this plan is occasionally desirable, and the bee-keeper will now and then find a case where he can profitably bother with it, much cleaner and better methods are now generally used. By transferring in this way one finds it to be a sticky and very disagreeable job. The combs and bees are messed up; the queen is likely to be killed, and the colony lost as a result, and stings are likely to be plentiful.

It very frequently happens that the bee-keeper who wishes to increase his stock by purchase will find it necessary to take such colonies in such hives as are available, even though he would prefer to pay a higher price to get them in good hives. It is usually advisable to transfer early in the season, or at least with a good honey flow ahead. If it is undertaken late in the fall, there is danger that the bees will not be properly prepared for winter, and will be lost before spring.

If the work is done at the beginning of the season, when the queen is active, it is an easy task to let the bees transfer themselves gradually. If the colony is in a box or keg, it should be turned upside down with bottom removed. All combs not containing brood which can easily be removed should be taken out. A new hive containing drawn combs, if they be available, should then be placed on top of the colony. If drawn combs are not to be had, full sheets of foundation should be used. It is a good plan to remove a comb containing brood from a strong colony, shake off the bees, and replace the comb with a frame containing foundation. The comb taken away can then be placed in the hive which is to be placed over the colony to be transferred. Care should be taken to see that no entrance is open in the box or keg, but that the bees must enter by way of the new hive. The bees seem to have an aversion to leaving honey and brood below the entrance, and if conditions are right they will soon move upstairs. Three weeks must elapse after the queen begins laying above to allow time for all brood to hatch, when the box hive may be taken away. If honey still remains it can be extracted and the combs rendered into wax.

When one transfers by the old method of cutting and fitting, usually a part of the combs will have to be discarded after the colony is successfully transferred, because of too much drone comb, crooked, or otherwise unsuitable combs. By this later method of gradual transfer, the bees are moved with little disturbance and no muss. The old combs are valuable for little but the wax they contain, and that is all saved.

Another Plan.—Some bee-keepers practice the method of drumming the bees up from the old hive into the new one above. When, after a few minutes pounding on the hive with sticks, most of the bees, including the queen, have gone above, the new hive is placed on the old stand and the old hive taken away. At the end of three weeks, when all brood has hatched, the young bees are united with the old colony and the old hive destroyed. Even though the old hive be left in place under the new one while the bees are moving upstairs, it is a good plan to drum them above to begin with, and then place a queen excluder under the new hive to prevent the queen from going down again.

Transferring from Buildings, Trees, Etc.—Nearly every bee-keeper of experience has been called on to remove a colony of bees from the side of some dwelling house, where they had found entrance through a crevice. Instead of tearing off a lot of boards and possibly injuring the building, one should begin by closing up all possible openings except one. Over this should be placed a bee-escape, either a Porter escape or a long wire cone, through which the bees can come out but cannot find the way back. A hive containing full combs or sheets of foundation should then be brought and placed with the entrance as near as possible to the escape (Fig. 18). In the hive should be placed

Fig. 18.—Transferring from hollow tree without cutting the tree.

one frame of brood, care being used to insure that eggs and hatching larvæ are both present. The bees coming out and unable to find their way back will enter this hive. Within a few days the bees will nearly all be in the hive, and the young bees emerging inside will shortly follow their fellows outside, to be barred from returning. As a result the entire colony with the exception of a very few bees will be in the new hive. The bees in the hive finding themselves without a queen will shortly raise one from the young larvæ in the comb provided. The old queen remains behind and the nurse bees in the meantime have cared for the brood in the old home, with the result that the colony has been transferred with little loss. After three or four weeks, when the bees are nicely settled in the new hive and the young queen has begun to lay, the escape can be removed, thus allowing the bees free access to the old brood nest. They will at once proceed to carry the honey into the new hive so that all of value to be left will be the wax, which, of course, cannot be obtained without opening the cavity. When everything is in place in the new hive, the bees can be moved to the desired location and the entrance to the house closed to prevent the place being occupied by another swarm. It will be necessary to use the usual precaution to prevent the bees from returning to the old location when moved.

QUESTIONS

1. Why is it advisable to start bee-keeping on a small scale?
2. Note the dangers to be avoided by the beginner.
3. Discuss the different kinds and sizes of hives.
4. What tools are essential?
5. Describe the essentials of a good cover.
6. Have you used an observation hive?
7. Discuss necessary considerations in buying bees.
8. Describe the bee's method of marking location.
9. What difficulties are to be met in moving bees and how can they be overcome?
10. How can one tell strong colonies?
11. What conditions should one look for within the hive?
12. Outline the best methods of transferring.

CHAPTER IV
ARRANGEMENT OF THE APIARY

The location of the apiary as regards shade, windbreaks, and convenience is very important. The questions of windbreaks and shelter will be considered more particularly in the chapter on Wintering. A suitable shelter from the prevailing winds of early spring is of great importance, and the reader is referred to Chapter XIII for a consideration of this phase of the subject.

If one is starting from the beginning and can plan accordingly, he will be able to so place the hives as to provide both a suitable situation and a convenient one. To keep down expenses is one of the essentials of successful bee-keeping, and to do so every operation should be performed with a minimum of labor. If one will but take the time to visit several bee-keepers, he will find that some have the apiary so arranged as to enable them to do the work with little more than half the labor necessary in others. Neatness and a fine appearance are desirable, but they are of secondary importance to convenience. If possible, the apiary should be on slightly higher ground than the honey house, and as near as possible. This will enable the operator to wheel the heavy loads of supers to the house with a minimum of effort.

Various plans of moving the honey from the hive to the honey house have been recommended, but a common garden wheelbarrow is perhaps as good as any. It is inexpensive, and can be pushed over rough and uneven ground easier than a cart.

Shade.—Bee-keepers are not fully agreed as to the value of shade in the apiary (Fig. 19). Something depends, perhaps, on the locality and the temperature during the heated season. While no shade is necessary in the spring and fall, or in early morning or late afternoon, shade during the heat of the day in mid-summer is very desirable. Comparisons of returns from colonies in the shade and in the open sun fail to show decided

SHADE

results that can be ascribed to shade or to the lack of it. For the comfort of the operator, if for no other reason, a shaded situation would seem to be desirable. Large trees should be avoided, if possible, because of the tendency of swarms to cluster so high as to make it very inconvenient to hive them. Many bee-keepers who have apiaries in the open provide shade boards made of cheap lumber obtained from dry goods boxes or similar sources. About two by three feet is the most popular size for such a shade board.

Fig. 19.—An apiary without shade.

This permits the sun to reach the body of the hive early and late in the day, while effective shade is obtained during the hottest period.

The shade of fruit trees, especially cherry and apple trees, if not set too close together is very satisfactory. Hives should never be painted with dark colors, because of the tendency of such colors to absorb heat. In extremely warm weather, combs will sometimes melt down and the colony be greatly injured or even destroyed. This sometimes happens even in the shade, especially if the hive be not well ventilated. The writer recalls an instance where several colonies in new and nicely painted hives met with this misfortune, while others in old hives full of cracks suffered no injury even though they were in the open sun and the unfortunate ones in the shade. It is desirable that the bees be so situated that there is always free circulation of air among the hives in warm weather. Large entrances greatly

Fig. 20.—A well-arranged apiary in California.

assist, and in very warm weather lifting the cover an inch and placing a block under it will also be of much help. When the bees begin to cluster on the outside, it is usually from lack of room to store or from lack of ventilation. In either case the need should be supplied to prevent loafing or untimely swarming.

Spacing the Hives.—It is a common practice to set the hives close together in long rows. This plan is not to be commended, because of the danger of queens entering the wrong hive and being destroyed. Neither is this plan satisfactory to the attendant, for in manipulating one colony he too often disturbs others near at hand, with the result that angry bees become very annoying. The preferred arrangement is to set the colonies in pairs. At least three feet of space should be between the pairs of hives (Fig. 20). The two may set within a few inches of each other. Some bee-keepers set them in fours, with two colonies with entrances facing to the east and two to the west.

Fig. 21.—A hive-stand of cement for two colonies.

A south front is to be preferred, especially in cool weather of spring and fall.

Hive Stands.—Hives placed directly on the ground do not give satisfaction for several reasons; the bottom boards soon rot and have to be replaced, grass grows up about the entrance and interferes with the flight of the bees, and it is hard to keep the hive level. Many kinds of hive stands are in use. Some use short pieces of board laid flat on the ground. These last but a short time and must soon be replaced. Four bricks make a good stand, if a piece of board is leaned against the front of the hive bottom for incoming bees to run up on. Four round bottles are sometimes used, one under each corner of the hive with the neck pressed into the ground. In dry weather this does very well, but in wet weather one corner or another is likely to settle, with the result that the apiarist must frequently go to the trouble of levelling them up.

If the apiary is placed in a permanent position, so that one

Fig. 22.—A tub of water covered with chipped cork makes a safe watering place.

does not need to consider the necessity of moving, concrete hive stands are, perhaps, the most satisfactory (Fig. 21). They are a little more expensive to begin with, but they are permanent, and once properly placed will remain in position indefinitely. They should be so placed as to leave the hive exactly level sidewise, but with the entrance slightly lower than the back to permit surplus moisture to run off readily. The concrete should extend several inches in front of the hive to prevent vegetation from growing too close to the entrance. If colonies are wintered out of doors, it is well to have these stands large enough to accommodate two or four colonies, whichever is the unit used for a single packing case. (See Chapter XIII on Wintering.) They can thus be left in the same position all year.

Bees coming in heavily laden during the honey flow often drop to the ground some distance from the hive and are unable to rise again. It is thus of considerable advantage to have the hive in such a position that they can crawl into it. For this reason high stands of any kind are not to be recommended. Any one who will watch the bees carefully for a few minutes during the height of the season will see at once that the loss of these heavily laden bees would be considerable in hives placed a few inches above the ground. For this same reason the hive stand is usually made with a gradual slope in front of the hive, to make it as easy as possible for the bees to reach home.

It is also important to keep down grass and weeds in the apiary. While considerable work is required to keep the grass closely cut during the busy season, it will pay well to do so. If the apiary is properly fenced, a few sheep will do the work in a very satisfactory manner without disturbing the bees. Ducks are sometimes used for the same purpose.

Watering Places.—In early spring when brood rearing is at its height, there is frequently much annoyance from bees about watering troughs, drinking fountains and other similar places. Large numbers of bees about a watering place frequently lead to an attempt by the town council to prohibit the keeping of bees within the corporate limits. It is in the small town, which has not yet reached the point of providing a common city water supply, that such difficulties most frequently develop.

Bee-keepers should bear in mind the need of the bees for large quantities of water for brood rearing, and see that it is within easy reach. In early spring when the weather is very changeable, it is important to save the bees as far as possible. If they are compelled to go far from the hive for water, many will be lost from the sudden drops of temperature common to that season. If there be a small spring, pond, or other open water near at hand, the bee-keeper need give the matter no further thought. If, on the other hand, the only available supply is from his neighbor's watering troughs, he is likely to save friction by providing an abundant supply near at hand. This should be attended to very early in the spring, and the supply constantly replenished before the bees form the habit of seeking it elsewhere.

Fig. 23.—In the foreground is a long trough with burlap lining for watering the bees.

There are many little plans that serve very well. A common way is to set out two or three tubs or half barrels and fill them full of water (Fig. 22). A quantity of cork chips is scattered over the top of the water to prevent the bees from drowning. Fig. 23 shows one of the best plans. In the foreground of the picture will be seen a long trough. This is made by nailing two six-inch boards together in a V shape and closing the ends, like an old-fashioned pig trough. The trough is lined with burlap to furnish a foothold for the bees. While fewer bees will be lost from this plan than most others, it has the disadvantage of requiring refilling frequently, as the large surface of water exposed results in rapid evaporation. On the other hand, large numbers of bees can get water at one time without crowding.

Hive Numbers and Records.—Some system of records is very helpful, and many specialists regard it as indispensable to best results. Various plans are in use, but the elaborate systems should be avoided. Unless one can make necessary notes while at work, there is danger that they will not be made at all. While a book record is best, it is difficult to make records of this kind when one's fingers are messed up with propolis, and the wind is blowing the pages. The principal advantage of the book record lies in its permanence, and the ease with which one can refer to previous notes.

Hive Marks.—In a large apiary some system of marking the condition of the colonies is necessary. When one examines the bees he will find some colonies weak, some strong, some needing more room, some with failing queens, some preparing to swarm, some queenless, and many other conditions. Where immediate attention is necessary, it is likely to be given at the time, but if something necessitates examination again after a few days have elapsed, some simple mark is necessary.

Stakes, etc.—Some bee-keepers use a quantity of stakes, and by setting them in different positions about the hive indicate the condition of the colony. F. W. Hall uses pegs or stakes for this purpose. A red topped peg indicates disease. Such a peg set at the left side of the entrance indicates disease is suspicioned, and in front of the left side of the entrance, that disease is known to be present. When the colony is treated, the peg is moved to the center of the entrance and if, after a later examination, no disease is found, the peg is moved to the right of the entrance. After disease is known to be eradicated, all pegs are removed.

In the same manner pegs not painted or of another color are used for other purposes. Thus one peg indicates a fair laying queen, two pegs a good queen, and three pegs a choice one. The positions of the pegs in front, at the side, or the rear of the hive indicate other conditions which it is desired to note. The advan-

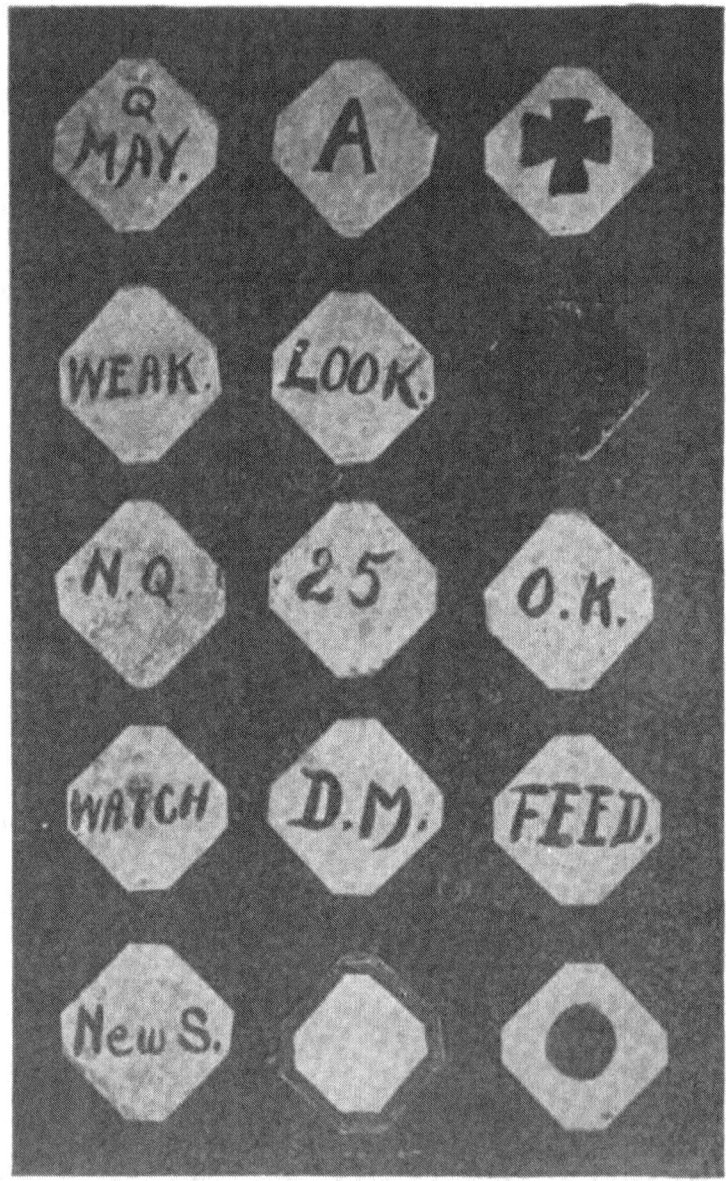

Fig. 24.—The Bonney hive-markers.

tage of this system lies in the fact that one can see at a glance the condition of each colony when passing through the yard.

Position of pegs is changed at each examination, to indicate the condition of the colony at the time.

Various modifications of this peg plan are in use. Some use a variety of colors, each color indicating some special thing, as red for disease, blue for queen, green for strength of colony, etc. The position of the stake tells the story. This plan is not entirely satisfactory, for the reason that so many stakes are needed, and they are not always sure to remain where placed.

Hive Markers.—One of the best things for this purpose is the Bonney hive marker shown in Fig. 24. The picture shows a variety of these markers. They are made of thin metal with sharp corners which fasten to the hive like a tobacco tag sticks to the plug. They will stay as long as desired, and yet are easy to remove. Consecutive numbers are provided, and in addition various signs or words to indicate anything desired. Feed, weak, watch, etc., are all suggestive of the condition of the colony. They can be had at such a low price, only a fraction of a cent each, that a quantity sufficient for the largest apiary can be had for a trifle. By using those made of zinc, any record may be made with lead pencil and will remain indefinitely. With a bunch of these markers in the tool box, and a lead pencil in the pocket, one is always prepared to make any necessary record at the proper time.

QUESTIONS

1. Note the things that are desirable in the arrangement of the apiary.
2. Discuss the matter of shade.
3. How should the hives be spaced?
4. Describe a desirable hive stand.
5. Under what conditions should water be provided?
6. Discuss various kinds of hive markers and systems of records.

CHAPTER V

SOURCES OF NECTAR

In taking up bee-keeping as a business, it is a matter of the utmost importance to select a location where suitable plants are available during as long a season as possible. The greater the variety of honey-producing plants the better. There is no single plant that can be depended upon to produce nectar in sufficient quantities every year. The ideal location is one where there is an abundance of willow, maple, dandelion, and fruit bloom early in spring, followed by white clover and sweet clover in abundance. This in turn should be supplemented with such plants as heartsease, sunflowers, golden rod, and asters for late forage.

There are many things to be considered in choosing a location, that will not be apparent at first glance. For instance, some plant may be present in quantity that is ordinarily considered as a profitable source of nectar, yet which for some unknown reason seldom yields in a particular locality. Alfalfa is a valuable plant for the apiarist under the conditions of the irrigated regions of the West, yet seldom secretes sufficient nectar to attract the bees in the moist sections east of the Missouri River. Buckwheat is rated as an important honey plant in New York, but is of little value in most Iowa localities. When the bulletin, "Bee-keeping in Iowa," which was published as No. 11 of the extension department of Iowa State College of Agriculture, was in preparation, correspondence with representative bee-keepers in all parts of that State brought only one report of buckwheat as a profitable source of nectar. Bee-keepers, reading of the wonderful crops of honey stored from buckwheat in some eastern States, might easily be misled into expecting similar results from this plant wherever a sufficient acreage was present.

Just what factors influence the secretion of nectar still remain to be determined. It is a well-known fact that some plants secrete very freely in some seasons, while in others with a large amount of bloom the bees will starve, or fare very poorly at best. Conditions that are favorable to the secretion of nectar with one plant seem to have the opposite effect on another. When white clover produces the heaviest flow in the Mississippi valley, alfalfa in adjoining fields will produce no nectar. Scientists are now studying the problems connected with nectar secretion, and it is hoped that the reasons for the great variation may shortly be better understood.

It accordingly becomes necessary for the bee-keeper not only to know the plants that furnish the raw material for honey production, but to be familiar with their behavior under the particular conditions with which he has to deal. Some years the honey crop will be good or bad over a large scope of country, while in others not more than five or six miles will be necessary to pass from a neighborhood where no honey is being stored to one where a profitable crop is gathered. The wide-awake bee-keeper can thus frequently, by moving his bees but a short distance, convert failure into success, and instead of having to feed his bees to get them safely through the winter, market a crop of honey.

One of the most successful bee-keepers of the Middle West has a location in the hills overlooking the Missouri River. His location is very desirable, for he has practically all important honey plants of that region within reach of some of his yards. His home yard is within easy reach of a large linden grove which furnishes some honey about two years in five. White clover in nearby pastures furnishes something about four years in five, and a good yield two or three seasons in five. Sweet clover, which is one of the surest honey plants, is also present in large quantity, and the bees also have a large area of Missouri River bottom land within reach. In this latter area they have access to large quantities of heartsease, wild sunflower, and other fall flowers. In such a location the chances of failure are reduced to the minimum, and seldom is there a year in which he does not get a surplus from some of these sources. In seasons when conditions are favorable for several important honey plants, he reaps a great harvest.

System Adapted to Honey Flow.—Upon the flora and conditions of secretion of nectar will depend the system of honey production which can be carried on most profitably. If the flows are short and very rapid, as is the case in many localities in the northeastern States, comb honey production can be carried on with very satisfactory results, and with profit to the producer. If, on the other hand, the flow lasts through a long season, and at no time is the honey coming in rapidly, it is very difficult to get a nicely finished article of comb honey. Not only are the sections likely to be poorly finished, but they will be travel-stained and unattractive in appearance. A small hive, while hardly to be recommended anywhere except in the hands of a comb honey specialist, should never be used in a region where the flows are not rapid. Locality then is really the first and most important point to be considered by the prospective bee-keeper. Not until he has settled upon his locality can he decide as to the system of management which he will follow, or the equipment which he will use.

Clover Region.—While there is a great diversity of local conditions, all the region from the Missouri River to Maine and south to the Gulf States, can be classed as the clover region. White clover, perhaps, stands at the head of the list of honey-producing plants in all this section. Alsike and sweet clover also are important. Basswood or linden, raspberry, buckwheat, and several other plants are important in various local sections, but the clovers are the main source of nectar throughout this vast region. Fruit bloom and dandelion are of great value throughout these States for spring brood rearing. In many places they are sufficiently plentiful to offer an important source of surplus, if the bees are ready for it. The bloom from these plants, however, comes so early in spring that the bees are usually not yet strong enough in numbers to make the best use of the nectar available. Coming so soon after the bees are first abroad after the long winter, the queens are stimulated to great activity and brood rearing begins in earnest. As a result, the hives are soon full of young bees, so that the colonies should be in the very best condition for the clover harvest.

Alfalfa Region.—The irrigated sections of the arid West may be classed as the alfalfa region. While much honey from other sources comes to market from west of the Missouri River, alfalfa is the main source of dependence. Sweet clover is rapidly extending its range in the same territory, so that it is also a very important source of nectar. The alfalfa plant seems to be at its best in the dry atmosphere of Colorado and surrounding States. Given plenty of water by irrigation, the results both in hay and in nectar are remarkable. Alfalfa may be said to be "king" in the Rocky Mountain States, all the way from Canada to the Mexican border.

From California we hear much of sage, orange, and beans as additional sources of honey production. Very little orange honey reaches the eastern markets, so that it can hardly be considered in speaking of the region as a whole. Sage, in years past, has been the source of large quantities of honey shipped east. Of late years, alfalfa, even in California as in other western States, is coming to be a very dependable source of supply.

The South.—In Texas are to be found many of the plants common to the eastern States, as well as some that are important in the arid West. In the irrigated sections of Texas alfalfa is an important honey plant. Here are also a number whose names are unfamiliar elsewhere, including huajilla, mesquit, and catsclaw.

In the southeastern section, beginning with the Carolinas, we still find the clovers and other plants common to the northern States mentioned as important in honey production. In addition, there are some peculiar to the South which rank even higher in the production of honey. Among the most important may be mentioned the gallberry and sourwood.

In Florida several other species of importance are brought to the bee-keeper's attention. They are not all confined to Florida, but may be found more or less abundantly throughout the Gulf States.

The list sounds strange to the bee-keeper of the North, for few of the names mentioned as important are familiar to his ears.

Tupelo or gum extends some distance to the north of Florida, but it is mentioned as one of the most valuable sources of nectar in that State. Palmetto and saw palmetto are peculiar to Florida.

The Florida honey flora is composed to a surprising extent of trees; magnolia, mangrove, titi, orange and many others are either trees or shrubs.

Honey Sources of Wide Distribution.—It is hardly within the scope of a work of this kind to consider in detail the resources of each section separately. There are, however, a large number of plants of wide distribution which are important yielders of honey or pollen, or both, over such wide areas as to merit further consideration.

During the height of the season, pollen is usually present in such quantities from so many plants, that those which yield pollen alone are of little interest. Those plants which yield pollen very early in the season, however, are second in importance only to the best honey sources. So valuable is an abundance of pollen early in spring for brood rearing, that it is very important that the apiary be within easy reach of pollen-bearing plants at this season.

Honey-dew is a secretion from small insects known as aphids. There has been much discussion concerning the origin of this product in the past, some holding that it was not only an insect secretion, but a plant secretion as well. The fact that drops of honey-dew are sometimes to be seen on the leaves of trees when no aphis is to be found probably gave rise to this impression. It is now quite generally agreed that honey-dew comes only from this family of insects as a secretion. Its presence where no insects are present is accounted for by the fact that quantities of the liquid are expelled by the insects from a higher point, and in falling it appears on the leaves on lower levels.

Several plants exude a sweet substance which the bees sometimes gather, but it is quite a different material from honey-dew. In seasons of scarcity of nectar the bees will seek any sweet material. At such times they are troublesome about cider mills, where they eagerly fill their honey sacs with the rich juice of the apple. They may even be found, at times, on decaying fruits which have been broken open. The saps from numerous plants when exposed by injury are freely sought. These substances are not honey-dew, though they are likely to be stored in the same manner.

The plant lice, or aphids, have a remarkable life history. The first generation of young to appear in spring is hatched from eggs and all are females. These in turn give birth to living young, but no males appear until several successive generations of living females have been brought forth. As the season advances males also appear and the cycle starts all over again.

There are many different species much sought for by ants. The solicitude of the ants toward the insects gave rise to the old story that ants keep cows. They do, in fact, seem to care for the plant lice for the purpose of securing the honey-dew which they secrete. The ants also use the bodies of the plant lice as food. It will thus be seen that the comparison of the ants and their herds is not so far wrong, for while the liquid secretion may be called milk, the ants may also be said to secure meat by the consumption of the plant lice themselves. With some species the ants are even believed to go so far as to carry the eggs of the plant lice down into their own nests to be cared for during the winter months, and to place the newly hatched aphids in position on their food plants in early spring.

Ants are not alone in their fondness for honey-dew, even though they may be alone in caring for the plant lice. Many other insects, including bees and wasps, are attracted to the feast when the product is abundant.

At times the plant lice become so abundant on the leaves of the various trees on which they feed as to prove disastrous to the tree. Plum trees are especially liable to injury from these insects. When the leaves begin to curl in early spring, it is usually a sign that plant lice of some kind are present. Hundreds of them will be found under one leaf.

It is usually from such forest trees as elm, hickory, and oak that honey-dew comes in sufficient quantities to be apparent in the hive. It is only in an occasional season that the bees gather honey-dew in noticeable quantities. At times it will fairly drip from the trees, and on such occasions, if no honey is coming in, the bees will work with tremendous energy in storing the substitute.

There seems to be considerable variation in the quality of the honey-dew honey, but as a rule it is not of good quality, and a bee-keeper should be very careful that it does not spoil his market for good honey. It is especially disastrous to the bees as winter food, and should never be left in the hive for winter stores, in the northern States, where the bees are confined for long periods without a flight. Where the bees are free to fly every few days during winter, the bad effect is not so apparent. (See Chapter XIII on Wintering.)

Sources of Early Pollen.—While some pollen is gathered from the early spring flowers, the most important sources are the forest trees. The elm is especially valuable, as it yields pollen in enormous quantities. When the elm trees bloom, the bees fairly cover them until the humming reminds one of the swarming season. At about this same season the maples (Acer) bloom (Figs. 25 and 27). These trees furnish not only pollen, but nectar also, and are a valuable source of supply at this season. The willows (Salix) likewise furnish pollen in abundance and nectar beside (Figs. 25 and 26). At times the bees are able to store some honey from the two last named sources.

THE DANDELION

The dandelion, or blowball, is widely spread over the temperate regions of North America, Europe, and Asia. It blooms early in spring, not far from the time of fruit bloom, and sup-

Fig. 25.—Soft maple and pussy willow are sources of early pollen and nectar.

plies both pollen and nectar in large quantities. While the nectar secretion does not seem to be constant, in an occasional season the bees will store surplus comb honey from this source. The pollen never fails when the plant blooms, and the bees fairly revel in it during its season of blossom. It comes just at the season when the need of food for brood rearing is greatest, and is consequently one of the most valuable. It is usually regarded as a

Fig. 26.—Catkins of pussy willow.

weed, and is especially obnoxious on the lawns. The bright yellow flowers are not unattractive, and if it was as difficult to grow as some of the cultivated plants, it would be much in

FRUIT BLOOM 55

demand. It seems to be human nature to despise the common.

Fruit Bloom.—The honey producer whose apiary is situated near large orchards is fortunate. If such tree fruits as cherries, plums, and apples all be present, the season of bloom will last

Fig. 27.—Blossoms of soft maple.

for several days, much to his advantage. A single variety will offer pasturage for only a very limited time, but it is abundant while it lasts. Most commercial orchards consist of several varieties, so that the season of bloom may last from one week

Fig. 28.—Fruit blossoms furnish large quantities of honey for early brood rearing. (Sear's Productive Orcharding.)

to three or four weeks, depending upon the kinds of trees available (Fig. 28).

The pear tree secretes nectar freely, and is one of the most valuable of the fruits. The peach also is of some value.

While the fruit trees bloom in profusion, the principal value of this nectar lies in the stimulation of brood rearing, because of the season in which it comes.

In Florida and California, the orange is of considerable value as a honey plant. It blooms usually in February and March, and lasts from twenty to thirty days. If conditions are favorable for nectar secretion, considerable surplus will be stored from this source, but it is not dependable.

Of the wild fruits, hawthorne, wild crab, and several others are very similar to the cultivated fruits in nectar secretion. The wild cherry is a large forest tree that furnishes considerable honey.

Besides the tree fruits, the bush fruits are of considerable value. The wild raspberry of Michigan is one of the more important sources of honey in that State. Its period of bloom is long and the honey of the finest quality. The plants grow on cut-over timber land that is very poor, so that a good raspberry location is quite likely to be permanent. A large acreage of cultivated raspberries is equally desirable and often to be found in the truck-growing sections near the large cities. The blackberry is also a valuable plant, although probably nowhere equal to a similar acreage of raspberry.

Currants and gooseberries are eagerly sought by the bees, and if present in sufficiently large acreage would be desirable pasturage.

Truck Crops.—In the vicinity of market gardens, the beekeeper often receives considerable benefit from the large acreage of cucumbers, which produces considerable honey. It is said to be of inferior flavor.

Carrots, cabbage, mustard, turnips, pumpkins, squash and several other cultivated vegetables add to the total production of the hive.

The California apiarists of some sections report valuable honey crops from the lima beans which are raised in large acreage in that State.

The Clovers.—In the markets of the world, honey from this family of plants stands supreme, both in quantity and quality. The combs are capped white, so that the product is of fine appearance and the quality is of the best. Honey from either white, alsike, sweet or other clover, or from alfalfa, is sure of a market at a fair price in almost any season. When the markets are glutted, the clover honeys are among the first to move, so that the apiarist seldom need fear being unable to dispose of his product. While there are seasons of short secretion, the clovers are, perhaps, as nearly sure to yield as any plants of wide distribution.

White Clover (Trifolium repens).—The most valuable honey plant in America. It ranges from Canada to the Gulf, and from the Atlantic west to Nebraska and Texas. Reaches its greatest value as a honey producer in the northern States. Perennial, with somewhat creeping stems. A fine pasture plant, common along roadsides and in pastures everywhere.

Alsike Clover (T. hybridum).—Alsike, or Swedish clover, resembles white clover in some respects, although much larger and better suited for culture as a forage crop. It yields honey freely of about the same quality as white clover. This plant succeeds on land where red clover will not do well, and when sown with a mixture of other grasses makes a very good meadow. When sown with timothy and red clover the resulting hay crop is much heavier than where timothy and red clover are sown alone.

Red Clover (T. Pratense).—Red clover would be a magnificent honey plant if the bees were only able to reach the nectar. The corolla tubes are too long for the length of the honey-bee's tongue. Occasionally a case is reported when a crop is supposed to be gathered from this source. If so, conditions must either serve greatly to reduce the length of the corolla tubes, or the nectar must be so abundant as to fill the tubes to a height within reach of the insects. Possibly the tubes are punctured by other insects. It is usually in extremely dry seasons that red clover honey is reported, and it may be that the unfavorable conditions may serve to dwarf the plant to some extent.

Although red clover produces nectar in great abundance, it can hardly be regarded as a honey plant because of the inability of the bees to gather it.

Mammoth, or Pea-vine clover, is another form or variety of red clover of a coarser or ranker growth. It is of little value for the same reason ascribed to the other, although bees may work on it slightly more than on the medium red variety.

White Sweet Clover (Melilotus alba).—Sweet clover is one of the best honey plants, and is very rapidly spreading into all parts of the United States. It bids fair very shortly to outstrip white clover in its total production, because of its wider range. It thrives on the rich lands of Iowa and Illinois, along the irrigation ditches of the arid West, on the heavy soils of the South, and the rocky farms of the East. It seems to thrive under a greater variety of conditions and to flourish under greater adversity than any other valuable forage plant. Farmers on high-priced lands of the corn belt are beginning to grow it in preference to red clover, or timothy, for hay, while the farmers on worn soils of the South and East are finding it to be a great soil builder, as well as a profitable farm crop. The quantity of forage produced from a given area is second to no other forage plant, and the quality, if properly handled, is excellent. Even in sections where it is regarded as a weed and consequently is not encouraged, it is gradually taking possession of waste places along railroads, highways, and unoccupied lands generally.

White sweet clover, also called Bokhara clover, is a biennial, seeding freely, and establishing itself readily under apparently unfavorable conditions.

The honey is light in color and spicy in flavor. By itself it is a little strong for some palates, but when mixed with other honey is of very fine quality. Sweet clover will one day stand at the head of the list of honey plants of the world, if the present rate of spreading continues. Large quantities of honey from this source are now reaching the markets from Colorado, Idaho, and other western States.

In Iowa one farmer, Frank Coverdale at Delmar, had nearly 200 acres of this crop on his farm. Sweet clover was the principal crop grown, and everything was planned to utilize it to the best advantage. Cattle, hogs and other stock were kept to consume the hay. Bees also were kept to gather as much of the nectar as possible. In 1913 more than a carload of fine comb honey was produced from the 300 colonies of bees on the Coverdale farm.

Yellow Sweet Clover (Melilotus officinalis).—The yellow variety of sweet clover is not nearly so widely spread as the white, and is not of so much value as forage. The honey yield, however, is good, and it is valuable as a honey-producing plant.

Alfalfa (Medicago sativa).—Alfalfa, or lucerne, is coming to be widely cultivated as a forage plant. It does not thrive to any extent except under cultivation. It is at its best in the irrigated regions of the West, where it is grown in very large acreage for hay and for seed. Under western conditions, it is a very valuable forage plant, yielding large quantities of fine honey. It seems to be of little value for bee pasturage in moist regions of the eastern States. Although blooming freely, it does not seem to secrete nectar, except in rare instances, and seldom produces seed in any quantity except in seasons of extreme drouth, when the bees will seek it freely for a time.

Basswood or Linden (*Tilia americana*).—The basswood, known as whitewood, linden, or limetree, is widely disseminated in eastern North America, being found from New England to Florida and Texas. It has also been introduced into California in a few localities.

In times past basswood was a very important source of honey, but of late years the linden forests are being rapidly cut off, and the land turned into farms or pastures. Wherever it is to be found in quantity, it is valuable for bee pasture, and in some seasons produces large quantities of nectar. It is not a dependable source, for it does not secrete nectar freely except in occasional seasons. When conditions are favorable, it offers about ten days of the finest honey flow possible. Some years immense crops are stored from basswood, so that the bee-keeper who is within reach of a considerable acreage of this forest can expect great benefit every third or fourth year, with a splendid crop once in ten or twelve years. The tree is a rapid grower, and will begin to bloom freely after six or eight years.

The wood is white, and much desired for making sections for comb honey. It is also utilized for making packing boxes of various kinds, some kinds of furniture, and for making paper.

Buckwheat (*Fagopyrum*).—In parts of New York, Pennsylvania, and the New England States where buckwheat is raised in large quantities, it is a very valuable honey plant. In some sections several hundred colonies of bees are kept in one yard, with buckwheat as the principal source of honey. Climatic conditions of the eastern States seem especially favorable to nectar secretion, and there it is very dependable, yielding some honey nearly every year. In the Central West it is seldom of much value for bee pasture, and yields only rarely. It is reported as of little value in Texas, except to bridge over a time when little else is blooming. In California there is another plant called wild buckwheat which is said to be of considerable value as a honey plant.

Buckwheat honey is dark, of a heavy body and strong flavor. Those who are accustomed to it often prefer it to milder flavored honeys, but in western markets it moves slowly, and at a lower price than the white honeys.

In the East it is the source of very large crops in some seasons, probably an average of fifty or more pounds per colony being secured from this source alone, under favorable conditions.

Golden Rod (*Solidago*).—The golden rods are of wide distribution, some species probably being found in every State in the

Union, as well as Canada and Mexico (Fig. 29). They are perennial herbs, blooming in late summer and autumn, mostly

Fig. 29.—The golden rod is an important source of fall nectar in some localities.

with bright yellow flowers. At least eighty recognized species are recorded.

The golden rod is an important source of nectar in many

sections. Reports of good honey crops from this source alone are frequently received from the eastern States. In the Central West it is less frequently mentioned as a honey plant, in some sections the bees seeming to pay no attention to it. It is said to yield considerable honey in Texas in favorable seasons, and is of some value, also, in parts of California.

The honey is usually thick, and when well ripened of good quality. The attractive flowers are much sought for by many insects beside the bees. Beetles in large numbers, especially blister beetles, frequent the blossoms.

Coming so late in summer, it is especially valuable in localities where the secretion is sufficiently abundant.

Wild Sunflower (*Helianthus*).—The wild sunflower is another summer and fall flower of wide range. There are many species, some of which may be found from the Atlantic coast to California and from Canada to the Gulf. They are tall, coarse weeds with bright yellow flowers (Fig. 33). Large numbers of insects of many species visit the blossoms of the sunflowers in search of the nectar.

Wherever these flowers are sufficiently abundant, they are the source of nectar. The cultivated sunflowers are of little if any value as honey plants, but produce seed in quantity, which is valuable as poultry feed.

The Jerusalem artichoke is a variety of sunflower, sometimes cultivated for the hogs. This plant grows wild in the upper Mississippi Valley States, and is regarded as a weed. It is frequently referred to as a valuable honey plant.

Many of the sunflowers are perennials, persisting for many years where once established. They are commonly to be seen along railroads, wagon roads, and on waste ground everywhere.

The honey is dark or amber in color.

Other Yellow Fall Flowers.—There are many coarse plants with yellow flowers that bloom in late summer and fall that add much to the sum total of honey stored. Coming at a season when, in many localities, there is no general flow, they are of considerable importance, even though the flow from them is never heavy. Such plants grow along railroads, roadsides, along streams, on

Fig. 30.—The cup-plant or rosin weed.

the margins of fields, and in waste places generally. Some of these plants are quite valuable, where sufficiently abundant.

OTHER YELLOW FALL FLOWERS 65

Cup-plant.—The cup-plant (*Silphium perfoliatum*), also called rosin weed, is a common roadside plant in the Middle West. The illustrations showing the flowers and also the plant give a

Fig. 31.—Blossoms of the cup-plant.

good idea of its appearance. By looking closely at Figs. 30 and 31 it will be seen that the stem is square, and that the leaves are grown together at the base, thus making a cup around the stem, from which the name is derived. These plants are abundant on rich lands along streams, and sometimes on uplands in the Mississippi Valley and eastward. They grow from four to eight feet tall, with numerous yellow flowers, so that where plentiful they furnish considerable pasturage for the bees, which visit them freely.

Crownbeard.—There are several species of crownbeard, some

Fig. 32.—The yellow crownbeard is much sought by the bees.

of which have white blossoms. Fig. 32 shows the common yellow-flowered variety which grows in open woodlands in the Middle West. The bees seek it very eagerly and a great humming may be heard about this plant when the bloom is at its height. The range of the different species of crownbeard (*Verbesina*), is from Pennsylvania to the Missouri River and south to Texas.

PARTRIDGE PEA 67

The list of such plants might be extended to great length, while no one of them is generally valuable the total bounty of all

Fig. 33.—Wild sunflowers are important honey producing plants over large areas.

is considerable, and each is important in limited localities where it is abundant.

Partridge Pea (*Cassia*).—The partridge pea is reported as an important source of honey in Georgia and Florida. Fig. 34

shows the common roadside species of the Middle West (*C. chamæcrista*), with blossom, seed pod, and leaf. The flowers are of an attractive yellow color, of about the size shown. This plant is very common along the sandy roads of the Middle West,

Fig. 34.—Blossom, seed pod and leaf of partridge pea.

where at times it may be found in abundance for miles at a stretch. Although bees visit the plant freely while in bloom, the amount of honey stored from this source is seldom noticeable in this region.

The plant is peculiar in that the nectar is not secreted by the flower proper, but by a gland at the base of the petiole. The season of flow lasts for several weeks in midsummer.

As it comes for the most part after the close of the honey harvest, the partridge pea in the northern States serves principally to keep the bees occupied until later flowers bloom in sufficient quantity to provide a real honey flow.

The quality of honey stored from this source is said to be poor.

Aster or Starwort.—There are said to be about 125 species of asters in North America, and also many species in Europe, Asia, and South America. These plants then must be familiar to the bee-keepers in all temperate regions of the world. Some species grow in open, shady woodlands, while others delight in the open sunlight of the prairies. They range in height from eighteen inches, or less, to eight feet. As a rule, the plants are many-flowered, as will be seen by Fig. 35. A plant with a small number of flowers was chosen in an attempt to secure greater detail. Sometimes hundreds of blossoms occur on one plant.

They range in color from white to blue and dark purple. Blue is perhaps the most common color. They have a tendency in some cases to become weeds, but are easily destroyed by cultivation, and are not often regarded as serious. The bloom in the northern Mississippi Valley States comes very late, lasting until killing frosts. In some years bees are found still working on these plants in November.

The asters, like the golden rods, are attractive to a large variety of insect life, many different species seeking them in addition to the bees.

The white-rayed flowers are said to be the most honey producers, some species, apparently, not yielding any nectar. The value of asters as honey plants is a little uncertain. While they yield considerable surplus in many localities, the honey makes very poor winter stores, and many reports show heavy losses from wintering bees on aster honey. The honey is said to be white with a mild flavor. In most localities it is mixed with dark fall honey, so that it is not often stored separately.

It is said to be rather thin, and by itself not to thicken up readily.

Boneset or White Snakeroot.—There are several closely

Fig. 35.—Aster honey makes poor winter stores.

related species of this plant (*Eupatorium*) known by the names of boneset, thoroughwort, and white snakeroot. The common species ranges from New Brunswick to Dakota, and south to the Gulf of Mexico (Fig. 36). Boneset is frequently spoken of as a honey plant. It blooms in late summer, sometimes persisting until frost. The plant is a perennial, and if left undisturbed remains for many years in open woodlands that are not too closely pastured.

Fig. 36.—Boneset or white snakeroot.

The species most common in the upper Mississippi region is known as white snakeroot (*E. urticæfolium*), which is supposed to be poisonous, and is commonly reported to cause the disease known as trembles in animals. Although much of this plant grows in the author's wild garden (Fig. 37), and also about the grounds where it is frequently eaten by the family cow, no bad effect has ever been noticed.

Milk sickness is said to be caused by the use of meat, milk,

FIG. 37.—Masses of white snakeroot in the author's wild garden.

butter or cheese from animals afflicted with trembles, so that snakeroot is popularly supposed to be the indirect cause of milk sickness in the human race, as well as trembles in animals.

In his book on poisonous plants, Dr. L. H. Pammel cites a number of cases where the disease, trembles, had supposedly been produced in animals by feeding them the extract of this plant.

Investigations that seem to contradict this conclusion are also cited, so that the connection of this plant to these diseases seems doubtful.

Fig. 38.—Two species of heartsease or smartweed.

The boneset of commerce is made from *E. perfoliatum,* which also is most often spoken of as a source of honey. The drug is well and widely known as a remedy.

These plants are an important source of fall honey.

Heartsease (*Polygonum*).—There are several species of plants belonging to this family (Fig. 38) that are variously known as smartweed, knotweed, heartsease, lady's thumb, water pepper, doorweed, etc. *Polygonum persicaria* or lady's thumb is perhaps the best honey producer of them all. There are fifty or more species in the United States and Canada, a number of which produce some honey.

This plant is particularly valuable in wet seasons, when an excess of moisture prevents the usual cultivation of many fields, or when, because of abundant water supply, these plants spring up in corn fields and grain fields after cultivation has ceased. At such times, large quantities of honey are sometimes secured from this source.

While the plants range over a wide area, they are particularly valuable as honey producers in the States of Iowa, Illinois and eastern Kansas and Nebraska.

The period of bloom lasts from August until frost and the honey varies greatly in color and quality. Much of it is a light amber, of fair quality, while some is very dark and of inferior quality.

Horsemint (*Monarda*).—Horsemint is most frequently reported as a valuable honey plant from Texas and nearby States. In this section very large yields are occasionally reported from this source. There are several species (Fig. 39) ranging from Quebec and New England, west to Dakota, and south to Georgia and Texas.

The corolla tubes are very deep, and it would hardly be expected that the bees could reach the nectar. Three species are reported as yielding freely in Texas, *M. clinopodoides* according to Scholl being one of the best honey plants. *M. fistulosa*, commonly called wild bergamot, is common in many of the States, from New England to the Missouri River and south to Florida and Texas. While at times this plant does not seem

MILKWEED

attractive to the honey-bee, it is widely reported as a source of honey.

Milkweed (*Asclepias*).—The milkweeds, also called butterfly weeds and silkweeds, are widely distributed on both hemispheres. About eighty-five species are recorded. Although the

Fig. 39.—The horsemints are valuable over a large scope of country.

milkweeds secrete considerable nectar and in favorable seasons considerable honey is the result, they are not generally favored by the bee-keepers because of the fact that bees sometimes become entangled in the pollen masses and are lost as a result. Kenoyer, in his studies of the relation of wild bees to plant pollination, found that wasps frequently had these pollen masses clinging to their feet. While it sometimes happens that a considerable number of bees may be lost from this cause, it is hardly as serious as it has been pictured in many of the printed articles.

The milkweed is a really good honey plant, and where sufficiently abundant contributes a liberal portion to the prosperity

Fig. 40.—Catnip yields honey abundantly.

Fig. 41.—Figwort or Simpson's honey plant.

of the apiarist. In "Gleanings in Bee Culture," for July, 1912, Mr. George H. Kirkpatrick, of Michigan, reports a yield of 1320 pounds from eleven colonies in eleven days, gathered from milkweed. Any plant that will produce one hundred or more pounds of honey per colony, in such a short period of time, even under extraordinary conditions, is worthy of attention.

GOOD PLANTS OF RESTRICTED DISTRIBUTION

The honey is said to be light in color and of good quality.

Good Plants of Restricted Distribution.—Some of the best honey plants are restricted to a small area. The sages of California have produced enormous crops of honey which have been shipped to the eastern markets in large quantities, yet little sage

Fig. 42.—The Rocky mountain bee plant is a valuable honey producer in Colorado.

honey is secured elsewhere in the United States. The palmetto and saw palmetto of Florida are important in that State, but not found far removed from the southern half of that commonwealth. The logwood of Jamaica and the West Indies, and to some extent in Florida, is of little interest to the bee-keepers of other sections. Nearly every State has some honey plants that are not of general interest because of the restricted range.

78 SOURCES OF NECTAR

Valuable Plants That are Seldom Abundant.—There are a number of plants that secrete nectar freely, which would be exceedingly valuable if sufficiently abundant. Catnip, *nepeta cataria,* (Fig. 40) originally introduced from Europe has become

Fig. 43.—Blossoms of the button bush.

very widely naturalized in the United States. When in bloom it is eagerly sought by the bees. Figwort, or Simpson's honey plant, *Scrophularia marilandica* (Fig. 41) is another widely distributed plant on which the bees work freely. The button bush, *Cephalanthus occidentalis* (Fig. 43) is a bushy shrub that grows in wet lands. In a few localities along the Mississippi it is sufficiently common to yield some surplus honey, which is light in color and mild in flavor. The wild cucumber, *Echinocystis lobata* (Fig. 45) is another wet land plant common everywhere along streams from New England to Texas but abundant enough to yield noticeable quantities of nectar in few places.

Plants Principally Valuable for Pollen.—There are a large number of plants commonly visited by the bees for pollen, which

FIG. 44.—Buckwheat is highly regarded in the East. (Sear's Productive Orcharding).

produce no nectar. Others, like the willows and maples, are valuable for both. Inasmuch as pollen is absolutely essential for brood rearing, it is important that it be within reach at all seasons, as nearly as possible. In a few localities it is never sufficiently abundant throughout the breeding season for best results. Fortunately, in most places, pollen is within reach most of the growing months.

80 SOURCES OF NECTAR

Among the important pollen plants may be mentioned several of the forest trees, including the willows, elms and maples already mentioned. In addition the box elder, walnut, hickory, ash, beech, birch, chestnut and aspens may be mentioned.

Fig. 45.—Where sufficiently abundant, the wild cucumber is valuable.

Corn is one of the most abundant pollen plants of its season, and the bees sometimes work on the tassels so freely as to give rise to the impression that it is a honey producer. Plant lice are sometimes to be found on the corn plant, and it may be that honey-dew is sometimes gathered from this source.

The list might be extended indefinitely, but it is only necessary to mention a few of the common plants, such as roses, sorghum, hops and ragweed. Many brightly colored flowers produce pollen but no nectar.

Bitter and Poisonous Honeys.—There are several plants that yield honey of such a disagreeable taste that it is of no value, except to feed back to the bees. The bitterweed, or sneezeweed of the Ozark region, blooms after the close of the clover harvest, and a good crop of white clover honey is frequently spoiled by mixing with the bitter honey. In such a locality, it is important that the bee-keeper be familiar with the time of blooming of such plants, and remove all good honey from the hive before the bees begin to store from them. The author has found this honey to be so bitter as to be absolutely unpalatable. Scholl says of the bitterweed of Texas (*Helenium tenuifolium*): "Honey yield good in favorable seasons; pollen; honey golden yellow, heavy body but very bitter, as if 50 per cent quinine and some pepper were added. June to October."

The honey from snow-on-the-mountain (*Euphorbia marginati*) is said to be bitter and disagreeable, and possibly poisonous.

Pammel, in his " Manual of Poisonous Plants," cites a number of plants which are supposed to produce poisonous honey. Among them may be mentioned mountain laurel (*Kalmia latifolia*), which is said to be common in the mountains of Virginia and nearby States. The honey from rhododendron is said to be poisonous also.

In choosing a location it is desirable to avoid the sources of these undesirable honeys as much as possible, and if they are present in the locality where one is operating, to use care to prevent them from being stored in the same comb with honey of good quality.

Cases of poisoning from honey in New Jersey are described somewhat in detail. The honey is said to produce a pungent burning taste as soon as the comb has passed the lips. In fifteen or twenty minutes the patients are seized with nausea, abdominal pain and vomiting. This is soon followed by loss of consciousness, coldness of extremities, feebly acting heart, and complete collapse. No less than eight cases were cited from New Jersey in 1896 by Professor Kebler.

The poisonous honey is said to have been, " dark honey which had a light brown color and a nauseating odor, pungent taste, caused a burning sensation in the back of the mouth similar to that of aconite." The source of this particular honey is not given.

Overstocking.—The question of overstocking has perhaps been the cause of as much discussion among bee-keepers as any one question relating to the business of honey production. The number of colonies that a given locality will support profitably is one of the most difficult matters to determine. Some writers offer a general suggestion to limit the number in one apiary to 50 or 75. However, seasons and localities vary so widely that no safe rule can be laid down. In this matter, the experience of other men in other localities, even though they be but a few miles distant, is not of much value. A locality may support 200 or 300 colonies splendidly one season, when 50 would nearly starve the next.

About the best advice that can be offered is to begin with a moderate number and gradually increase until the average production per colony is no longer profitable, or rather until it would be more profitable to divide the bees into two separate yards three or four miles apart.

If one happens to have a location where fruit bloom is abundant for early brood rearing, followed by a large acreage of white and sweet clover, with a liberal supply of fall pasturage, he can keep a large number of colonies profitably in a single yard. As a general rule, the number of colonies kept in a single yard in the Central West is thought best not to exceed 100. However,

Frank Coverdale, at Delmar, Iowa, had 300 colonies in one yard. He formerly kept his bees in several outyards, in deference to the general idea that the locality could be easily overstocked. He found a great saving in time and expense in having the bees all at home, where they could be under constant supervision.

At Center Point, Iowa, S. W. Snyder, secretary of the Iowa Bee-Keeper's Association, kept more than 200 colonies in one yard. About two miles distant another bee-keeper had nearly 250 colonies in one yard. Thus there were nearly 500 colonies within two miles. In the town of Maquoketa, Iowa, there were several apiaries, some of which had more than 150 colonies each. Within a mile or two outside there were several more apiaries, thus bringing the total number of colonies much above that commonly thought to be profitable. Yet in all the above-mentioned cases the yields were very satisfactory.

A number of instances have been published where from 500 to 700 colonies have been kept in one yard, in the States of New York, Idaho and California.

Apparently, the number of honey-producing blossoms available for early brood rearing, and during the season of greatest dearth, have an important bearing on the number of colonies that can be supported through the year. During a heavy flow from any source, it seems at times that thousands of colonies could find support. If an insufficient supply of honey and pollen is available to support the colony during long periods of comparative idleness, the available stores will be too heavily drawn upon for support, and the number of colonies should be reduced for profit.

In many localities a few colonies will make a very good showing, when a substantial increase in the number will so reduce the average per colony that they are no longer profitable.

Prior Rights.—It is a common trait of human nature to crowd in where some one else has found a profitable opening of any kind. It very frequently happens that when a bee-keeper has become established in a locality that produces good yields, others will locate within a short distance of his apiaries, and the number of colonies brought in will so reduce the surplus secured that no one will get satisfactory returns. This is not only short-sighted business policy on the part of the newcomer, but very unjust as well. While a bee-keeper has no way to establish a legal right to the bee pasturage, it would seem that the first man on the ground should have some moral rights that should be respected. Indeed, there has come to be an unwritten law among bee-keepers that does respect the rights of the man already located. Unfortunately this unwritten law is not always recognized, and much friction sometimes develops as a result. The only remedy is to move to a new locality, or be patient until the newcomer will realize that there is not room enough for two, and move on in search of richer fields.

THE BEE AS A POLLENIZING AGENT

The value of the honey-bee in the pollenation of blossoms has come to be so generally recognized that commercial fruit growers and gardeners are anxious to secure the location of an apiary near their plantings. Since Darwin laid down the law that nature abhors self-fertilization, there has been much study of the problems of cross-fertilization and the agents that serve to accomplish nature's purpose in the distribution of pollen. While there are numerous butterflies, wasps, wild bees and other insects that assist in the work, the honey-bee, because of its greater abundance, and because it can be readily controlled, has come to be recognized as the most valuable agent for certain plants.

In this connection a quotation from Dr. Burton N. Gates, of the Massachusetts Agricultural College, will serve to show the present recognition of this fact by well-known authorities.[1]

The value of the honey-bee in cranberry cultivation has but recently been recognized. The cranberry industry of Massachusetts, for instance, is

---

[1] "The Value of Bees in Horticulture," by Burton N. Gates, in 3rd Annual Report, State Bee Inspector of Iowa, 1914.

worth between one million and a million and a half dollars annually. It has been observed that in certain years certain parts of the cranberry bogs fail. Dr. Franklin, at the experimental bog in Massachusetts, has carried out experiments, the details of which show that bees are of service and explain that the failure of bogs or parts of bogs may be attributed to the inability or lack of bees to work the blossoms while the vines are in bloom. It has been shown, too, that the inability of bees to visit these bogs was due to climatic conditions, the prevalence of winds or coldness in that part of the bog. With the large number of blossoms which are produced on cranberry vines, it was also established that bees maintained purposely for their service in pollenation were an insurance to cranberry growers who are now maintaining apiaries in proportion to the size of their bogs.

In Cucumber Growing.—The cucumber has been mentioned. In Massachusetts in recent years, cucumber growing under glass has developed. Originally the growers "fertilized the plants" by hand, a most laborious process. Bees were later introduced and found to be indispensable, especially in the larger commercial houses. One grower, for instance, has forty acres under glass. Taking the industry in Massachusetts as a whole, it requires between two and three thousand colonies of bees annually to serve in the cucumber greenhouses. These colonies are largely reduced by the extremely unfavorable conditions of greenhouse life, so that cucumber-growing-under-glass demands that the bee-keepers raise bees purposely for greenhouses.

A $3800 Crop Due to Bees.—I have in mind a specific instance reported by one of our Agricultural Experiment Stations. In one of the Western States there are two commercial apple orchards of about equal acreage, of similar location and age, each in a "pocket" in the foothills of an admirable fruit land, both well drained and protected from frost. One orchard bore heavily for successive years; in the other there was no crop, although the trees blossom heavily each spring. In despair of financial ruin, the owner called the assistance of a State Experiment Station. A pomologist and entomologist was sent, who examined critically all the conditions in each of the orchards. He was about to return without solving the problem of failure, when the question arose, were there ever bees maintained to set the orchard which had fruited? It was asserted, however, that neither orchard had ever had bees. However, the problem was not given up and the ground was again gone over. As the experiment station man was about to leave without finding any apparent reason for failure, he chanced to see a stream coming in one of the orchards from underneath a pile of swale. Further investigation revealed a fallen log, sunken in the damp land, sheltering a large colony of bees. It is needless to say in which orchard the log was. Immediately bees were secured for the failing orchard; the owner then netted $3800 on his crop.

Orchardists Realize the Value of Bees.—There has been a marked change in sentiment on the part of the fruit growers during the past few years, since they have come to realize the value of the bees in their orchards. Not many years ago frequent attempts were reported of trying to secure the removal of the bees from the vicinity of orchards, on the plea that the bees injured the fruit. Even yet cases are sometimes reported of the supposed injury of grapes by the bees. It has been so often demonstrated that the bees cannot injure sound fruit that there is no need to state the proof here. There are times, when the bees are finding no nectar, when they become very annoying by seeking the orchards and vineyards in search of the juice of fruits that have been injured by birds or other insects. At such times they become so troublesome that there is some ground for complaint, although they do no real injury to fruit.

The misunderstanding between the bee-keepers and fruit growers is very happily being cleared up, so that it is only now and then in the case of some fellow who is behind the times that trouble of this kind occurs.

On the other hand, horticulturists are loud in the praise of the honey-bee, and hundreds of testimonials as to her value in the fruit plantation could be cited. In an article " The Development of the Apple from the Flower," that recently appeared in " Better Fruit," O. M. Osborne, of the Horticultural Department of the State Normal School of Idaho, made the following statement:

> Without the aid of the bees but very little, if any, pollen would ever reach the stigma, for the pollen of the apple is a trifle sticky, and, unlike that of the corn tassel, ragweed, and several other familiar plants which are powdery, it cannot be distributed by the wind.

Since the horticultural authorities generally have come to realize the true place of the honey-bee in the orchard, old prejudices have quickly been broken down, with the result that progressive fruit growers, in many cases, are ready to offer some substantial inducement to the apiarist to locate near their plantations. Unfortunately, a few fruit men are still inclined to spray their trees while in bloom, greatly to the disadvantage of the bee-keeper. The horticultural authorities, here again, are coming to the rescue of the bees, and are showing wherein it is to the disadvantage of the fruit grower to spray during this period, because of possible injury to the fruit crop, as well as to the bees.

QUESTIONS

1. Discuss the importance of a good location.
2. Note the difference in nectar secretion of the same plant under different conditions.
3. How is the system of honey production affected by the locality?
4. Outline in general the clover region; the alfalfa region.
5. At what season are pollen-bearing plants most valuable and why?
6. Discuss honey-dew and its origin.
7. What are the principal sources of early pollen?
8. Of what value is fruit bloom to the honey producer?
9. What class of honey is highest in quality and highest in price?
10. Discuss the clover family and mention the varieties of greatest value.
11. In what sections is buckwheat of importance?
12. Name some trees that furnish honey in abundance.
13. What fall flowers add to the bee-keeper's revenue?
14. Discuss the plants restricted to small areas.
15. Mention some plants of value for pollen only.
16. Discuss bitter and poisonous honeys.
17. Note the problems of overstocking a locality.
18. Of what value is the honey-bee as a pollenizing agent?

CHAPTER VI
OCCUPANTS OF THE HIVE

One of the most fascinating studies in all creation is the social insect world. Bees, ants, wasps, and termites all have a good deal in common. The bee, because of its practical value, has attracted more attention than any of the others. Well it may, for the social life of the community is none the less interesting because of the fact that the honey and wax produced may be made to support the investigator in comfort while he is pursuing his studies.

This volume is designed to be a practical book, and so it would hardly be the place to go into detail, except in so far as the knowledge may be applied to practical purposes. Maeterlinck has indulged his fancy in dealing with this phase of the honeybee with the result that he has produced a most interesting story, based upon the specialized social life of the hive. There is much of truth as well as poetry in that wonderful book, which has perhaps been more widely read than any other volume ever written about the bee.

The Queen.—The life of the hive centers in the queen (Fig. 46), the mother of the community. Apparently her only duty is to produce eggs in large numbers, that the colony may be perpetuated, and that the inmates may be sufficiently numerous to enable them to store enough honey to meet the needs of adverse seasons in summer as well as the long period of cold, dreary winter. She rarely leaves the hive except on her mating trip and to accompany a swarm. Most of her life is spent in the hive, quietly attending to her duties. Apparently the egg from which she hatches is no different from any of the thousands of others that produce workers. The marvelous physical change that takes place when an egg is taken from a worker cell and reared in a queen cell is one of the most striking studies in the result of environment. Only sixteen days are required for the queen to reach maturity from the time the egg is laid, while the worker requires twenty-one. The queen is much longer in shape and looks to be one-third to one-half larger than the worker. The queen lacks the wax-secreting organs of the worker, while her own sexual organs are fully developed. She lacks the pollen

FIG. 46.—Queen laying in a newly made comb. The queen can be recognized by her greater length (see arrow) and the circle of attendants facing her.

baskets and brushes which are conspicuous in the worker. True enough she would have no possible use for any means of carrying pollen or secreting wax in her work of being a mother to a family of a few hundred thousand offspring during her lifetime. Nevertheless, as far as can be seen, the physical changes are entirely the result of a change of environment. The queen will remain in the hive, so her eyes are not nearly so well developed. She has no need to discover the distant fields of clover. Her life will be much longer as a mother than had she developed into a worker. The life of a queen may be from one to five years.

Queen cells somewhat resemble acorns in shape (Fig. 47). Three days pass from the time the egg is placed in the cell until it hatches into a tiny white larva. The little larva is provided

Fig. 47.—Natural-built queen cells.

with a liberal quantity of royal jelly on which it feeds. At the end of six days the larval growth is completed and the cell is sealed. Seven days are required to complete the transformations from a larva to a mature queen bee, and the cell is opened from within, and the queen appears upon the comb. Warmer or colder weather may slightly influence the period of development, so that it may be a little longer or a little shorter, but sixteen days is recognized as the normal period of development from the egg to maturity.

The instincts of the newly hatched queen are very different from those of the newly hatched worker. The worker mingles freely with her fellows without the slightest hostile action. The newly hatched queen begins at once to search for possible rivals. Should there be other unhatched queen cells, she will at once destroy them, if unmolested by the workers. Should the colony be preparing to swarm, the unhatched queens will be protected by a guard of workers. Ordinarily the needs of the colony are met by a single vigorous queen, and she promptly dispatches any others, either mature or in the cells. On one occasion the author observed three young queens to emerge almost simultaneously. They immediately gave battle, and but a few moments elapsed until they were in a death grip.

It sometimes happens that an old and failing queen will remain in the hive for a time with the daughter who will later supersede her. Apparently then there is no antagonism between them, for the mother in the very nature of things must shortly die. Just why there is such a change in the attitude toward each other in cases of this kind is hard to understand.

Usually when the queen is from five to seven days old, she departs on her wedding flight. The mating takes place in the air during the warm period of the day, when the drones fly in greatest numbers. The organs of the male are torn violently away, and carried back to the hive by the newly impregnated queen. The entire content of the male seminal fluid is absorbed by the queen, who retains it in a special sac, where it continues to fertilize the eggs during the life of the queen mother. It is no longer questioned among practical bee-keepers that the queen mates but once, and that one impregnation is sufficient for life.

One of the most remarkable things in the life of the bee is the fact that an impregnated queen may produce at will either male or female offspring, while the unimpregnated queen produces male offspring. Apparently, the eggs from which drones are hatched are never impregnated in either case. When Dzierzon first discovered this fact, which is called "parthenogenesis," his newly formulated theory of the ability of virgin mothers to produce male offspring was ridiculed as impossible. However, his observations were later confirmed by careful observers, and of late it is regarded as a settled fact, rather common among insects.

A vigorous queen will lay more than her own weight of eggs daily during the height of the season. Nurse bees wait upon

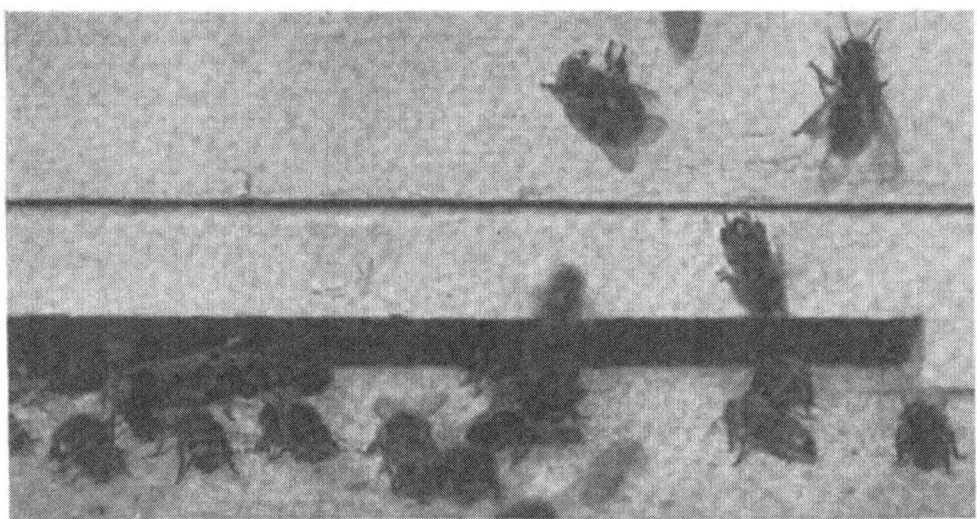

FIG. 48.—Worker bees at the entrance of the hive.

her constantly, and feed her freely with highly nutritious and ready digested food.

The Worker.—Upon the worker bees (Fig. 48) devolve all the labor of the hive. A lifetime of toil is their normal portion. Building the combs, gathering the nectar and pollen to furnish food for the community, secreting the wax, feeding the queen and drones as well, nursing the young bees, guarding the hive against robbers, and carrying out refuse to keep the home clean are a part of the manifold duties that they are called upon to perform.

By inheritance the worker is apparently in every way similar to the queen, but the difference in environment under which she develops, makes of her a very different creature. The practical apiarist takes advantage of this fact to utilize the eggs and larvæ from worker cells to rear queens in large numbers when such are desirable. The worker is reared in the ordinary cell in which honey is stored. The close confinement of the narrow cell deprives her of normal sexual development, and she is incapable of mating and of normal sex life. In addition to the larger cell occupied by the queen larvæ, the richer food, royal jelly, seems to have a great bearing on the difference in development.

It is now generally agreed that the newly emerged bees are first occupied with the duties within the hive, such as comb building and nursing of the young. Later they go to the fields to gather honey and pollen, and thus continue to the end of life. It is probable that under normal conditions the young workers do not go to the fields until they are from two to three weeks old. During the honey flow the average length of life among the workers is short, probably not much more than six weeks, while those hatched late in summer after the harvest is gathered may live until the following spring.

Twenty-one days is the usual period of development, from the time the egg is laid until the worker leaves the cell. Three days are required for the hatching of the egg, six days are spent in the larval period, and twelve days within the cocoon after the cell is sealed. This period varies slightly according to conditions of heat or cold, or possibly because of other abnormal conditions.

A newly emerged bee is easily recognized by her small size and velvety appearance. One is reminded of a baby just learning to walk, by the uncertain attitude of the youngster. On sunny afternoons large numbers of the young bees will be seen in flight about the hive. These play spells are often mistaken for evidence of robbing by the novice. When brood-rearing is at its height, a pint or more of young bees will be emerging every day to replace the old bees, which are wearing out in field work. When they are about a week old, they take a flight to try their wings and to learn something of the location of the hive and surroundings. The first trip to the honey field will not be taken until later.

Fertile Workers.—Occasionally, in a queenless colony, a worker will develop to the point of laying eggs. As she is incapable of mating, her offspring will all be drones, which are of no value to the hive and the colony will soon perish. Fertile workers lay here and there over the comb with little regularity. Sometimes several eggs will be found in the same cell, and the next cells will be empty. That the offspring are drones will be evidenced by the high arched cappings, like rifle bullets, which are peculiar to drone brood.

Much has been written about methods of saving colonies with fertile workers, but the productive bee-keeper will have little time to bother with them. As a rule, the best plan is to unite the colony at once with another which has a good queen and thus save what bees are left. If fertile workers are present, several are usually to be found in the same hive.

When the worker has served her purpose in life and can no longer render a service to the community, she will still persist in going to the field until she dies, or if she refuse to do so will be dragged from the hive in the most merciless manner by the busy sisters, whose only thought is for the prosperity of the community. With the social insects, such as the honey-bee, the community is everything, while the individual receives little consideration.

Under normal conditions, a colony of bees will consist of perhaps 20,000 workers, a single queen, and a few dozen or possibly a few hundred drones. The number of drones will depend largely upon the kind of bee-keeper in whose apiary they reside. A very prolific queen with plenty of room, and otherwise favorable conditions, will produce such large numbers of eggs that possibly 50,000 or more bees may be present at one time. Under unfavorable conditions the colony may be reduced down to the point where but a few hundred bees remain, and yet be revived with careful attention.

The Drone.—The sole purpose that the drone (Fig. 49) serves is the perpetuation of the species. As one mating is sufficient for the entire life period of the queen, except in rare instances when the first mating is not a complete one, not one drone in a hundred will ever have opportunity to serve the purpose for which nature designed him. The over-abundance of drones is a severe tax on the production of the hive. Nature provides for emergencies by producing large numbers of drones, to insure the presence of a male at the time and place of emergence of a virgin queen. In a large apiary, however, where many colonies are present, there is little danger but that this need will be met even though the bee-keeper take great care to reduce the production of drones to the minimum. Drones appear in the hive during the productive period of the summer. In April or May when brood rearing is active they will be seen

Fig. 49.—Drones.

and will continue about the apiary until the close of the honey flow in late fall, unless there is a dearth of nectar, when they will be summarily driven from the hive to perish.

The drones are reared in cells of the same shape as the worker cells (Fig. 50). They are, however, somewhat larger in size, and the cappings are raised like rifle bullets. These high arched cappings will show at a glance the presence of drone brood in the hive.

The practical bee-keeper reduces the available space for drone brood to the minimum by using full sheets of foundation in all brood frames. Where full sheets are used there will be but little drone comb built. A few cells here and there in the corners and along the bottom bars will provide sufficient space for the rearing of plenty of drones for all practical purposes. If the bees are allowed to build at will they are quite likely to build large quantities of drone comb. This is especially likely to occur in colonies where the queen is old or not very productive.

Fig. 50.—Combs showing queen cells and capped drone and worker brood.

A new colony with a vigorous queen will frequently fill the hive with worker comb, because the queen occupies the space as fast as built and fills every available cell with an egg. As soon as they have built sufficient for the queen's immediate needs, they are likely to begin to build drone comb. Apparently it is easier to construct and does not tax the bees quite as heavily.

In neglected apiaries it is not uncommon to find hives with from one-fourth to more than one-half of the comb surface composed of drone cells. This insures that a large number of drones will be reared during the active season. A surplus of drones who are loafers and consumers instead of producers will turn what might have been a profitable colony into a non-producer, or even one that will require feed to winter successfully.

These male bees consume quantities of stores, not only during the period of their development, but in the mature state as well. They are helpless fellows, not even able to feed themselves.

The drone is much larger and heavier in appearance than a worker. He has aptly been called a corpulent fellow. He has no sting and flies with a large buzz, which tends to frighten the novice who is unable to recognize his true character. The period of development is longer than that of either the queen or the worker. Three days is the period required for the egg to hatch as with the others. About seven days are spent in the larval period, and fourteen days elapse from the time the cells are sealed until the transformation is complete.

The act of copulation is fatal to the drone. As previously stated, the organs of generation remain attached to the queen for several hours, until the entire supply of seminal fluid enters the sac of the queen. After this is accomplished the attendant workers remove the parts from her.

The life term of the drone is very uncertain. If conditions are favorable he may live for several weeks, or maybe months, until by chance his life is terminated by meeting with a queen, or perhaps by accident. Otherwise he may live until the close of the honey harvest leads his provident sisters to accomplish his destruction. While the worker may sometimes sting the drones to death, it seems to be more often accomplished by simply refusing to provide them with food, and by driving them from the hive when they soon perish.

Drone traps are on the market to enable the bee-keeper to reduce the number of boarders in the hives. However, it is much better to prevent breeding them in the first place, as the food provided in rearing them together with the labor of the nurse bees is all lost. Combs composed largely of drone cells should either be used as extracting combs above a queen excluder or else rendered into wax and replaced with sheets of foundation. The productive bee-keeper can ill afford to divide his profits with useless drones.

RACES OF BEES

Italians.—While there are a considerable number of races of bees, those commonly known in this country are all that need be considered in a work of this kind. First and foremost among them may be mentioned the Italian, which is generally recognized as the most valuable under the conditions of this country. These bees have become so widely distributed in many parts of the country that together with their crosses, commonly spoken of as hybrids, they are about the only bees to be found in many localities.

There is considerable difference in the appearance of the various strains of Italians. The three banded strains are usually regarded as more desirable, although the goldens are highly regarded as well.

Pure Italians are usually very gentle, are more resistent of disease, especially European foul brood, than other races, and also repel the wax moth much more effectively.

These bees have been tried under so many conditions, by so many bee-keepers all over the country, that they may well be regarded as entitled to first place in popular esteem.

Cyprians.—For a time the Cyprians were quite the rage. They came from the island of Cyprus. They resemble the Italians, but are much more difficult to control. In fact they are so cross that most bee-keepers have discarded them, and queens of this race are seldom offered for sale.

Common Black or German Bees.—This seems to have been the original stock first imported into America, and which became common everywhere before other races were introduced. Accordingly, more or less of this stock is likely to be found in any locality. As above stated, the Italians have been so extensively cultivated in many regions that the blacks are no longer pure, but are only found mixed with Italians.

They are not nearly as gentle as the Italians. Neither do they resist disease or moths with much success. When the hive is opened, they rush here and there with such nervous haste as to be very disconcerting to the bee-keeper. The queens are very difficult to find, and taken altogether they are unsatisfactory bees to handle on a commercial scale.

Carniolans.—The Carniolans somewhat resemble the blacks in color, although the bands are more distinct. They are gentle like the Italians, and are quite popular in some localities. The principal objection to them is the excessive swarming propensity. They rear large quantities of brood as the queens are very prolific.

Caucasians.—These bees resemble the common blacks so closely that the novice will find it difficult to tell the one from the other. They are, however, said to be a gentle race, and have a few champions who assert that they are the best bees ever introduced.

It is a pretty safe rule in the average American locality to depend upon the Italian, unless some other race has been successfully tried in the neighborhood. It is only fair to say, however, that no other race has been tried under such widely different conditions as has the Italian. It is possible that with an equal opportunity to demonstrate their good qualities, either the Caucasian or Carniolan races may rival them for popular favor.

QUESTIONS

1. Describe the life history of the queen and note her peculiarities.
2. In what respect do the workers differ from the queen?
3. Discuss fertile workers.
4. Give the life history of the drone and tell something of his habits.
5. Note the difference in the three kinds of cells in which queens, drones and workers are reared.
6. Tell something of the different races of bees.

CHAPTER VII

INCREASE

One of the perplexing problems to the beginner is that of securing increase without loss of a honey crop. The control of natural swarming is probably the most difficult problem that the bee-keeper has to solve in the average locality. Certain plans will work all right for several years, until the bee-keeper begins to congratulate himself on having learned the secret, when suddenly they will swarm in spite of the best possible attention and once the swarming fever is on they are likely to keep it up until he is nearly beside himself.

Natural Swarming.—There has been much written about why bees swarm, and the control of conditions that lead to swarming. It should be remembered that with bees and other social insects the community is the unit, rather than the individual. The workers are incapable of reproduction, and accordingly no matter how great an increase there may be in their numbers in a hive, it is but temporary, and makes no permanent difference in perpetuation of the species. Swarming is then the expression of the instinct of procreation or increase.

Normally, the bees will swarm at about the height of the honey flow, when natural conditions favor the establishment of the new colony. As a rule, nearly enough honey will have been brought to the old hive to carry the colony through the winter, and at this season the new swarm will be able to establish itself with a minimum of danger. While the natural effect tends toward the safety of the bees, the practical effect to the bee-keeper is to divide his colonies at the time when greatest profit may accrue from large colonies, and results in increase of bees at the expense of the honey crop. The thing the bee-keeper should strive to do is to make his increase either before the honey flow begins or when it is nearly over, so that he will get both increase and a crop.

Certain conditions favor natural swarming, as, for instance, small hives that are soon filled with brood and honey, leaving the queen little room in which to lay, and the workers no place to store the incoming nectar. The old-time bee-keeper usually placed but one super on top of the hives and when that was full took it off and replaced it with another. As a matter of course, when the hive became crowded the bees began to hang out in large clusters for want of room, and the owner decided that they were preparing to swarm, which they usually did before many days. The practical apiarist will not tolerate this hanging out. He knows that, as a rule, either the bees are crowded for room, or there is not sufficient ventilation.

If on examination he finds an abundance of room for storage, he gives a larger entrance, or, if the weather is very hot, even lifts the hive off the bottom board a half inch or more and supports it on blocks at each corner.

A heavy honey flow seems to act as a check on swarming, and in localities where the honey flow comes on with a rush and continues heavy during the principal period of nectar secretion, there will be less difficulty in controlling swarming. In such localities, if the bees are furnished with plenty of room in which to store the honey, and the brood nest is large enough to permit the queen to continue her activities, the bees will apparently have no instinct but to gather honey as rapidly as possible. If the flow stops suddenly, there may even be little if any swarming. On the other hand, in most of the northern States, where there is a light flow from fruit bloom and dandelion in advance of the clover flow, the bees are likely to be swarming full tilt at about the beginning of the best flow.

Clipping the Queens.—It is a common practice among apiarists to clip the wings of the queens to prevent their escaping with the swarms. If the bee-keeper is constantly on hand this plan works very well. When the swarm issues and the air is full of bees, the bee-keeper goes to the hive from which they have issued and usually will have little trouble in finding the queen moving about in front of the hive. It is then an easy matter to place her safely in a cage, and to remove the old hive from which the swarm issued and put a new one ready for the swarm in its place. If the bees cluster in a convenient place they may be shaken into a basket and dumped in front of the new hive at once, and the queen released and allowed to run in with them. Usually, the bees will shortly miss the queen and return to the old location of their own accord, and when they begin to enter the hive the queen may be released. This is a very easy manner of hiving swarms when the owner is in the yard when they issue. If no one is present when the swarm comes out, even though they be found while still clustered, it will be difficult to find the place from which they came, in a large apiary, and the swarm is likely to return before the queen is found. Colonies that are not permitted to swarm naturally are likely to come out again with a young queen, with which they will make off to distant scenes.

Clipping is a decided advantage where large trees are near the apiary, as it is a difficult and unpleasant task to capture a swarm that has clustered in the top of some tall tree, perhaps forty feet from the ground.

There is another advantage in having clipped queens; one can tell the age of every queen in the yard if records are kept. If the queen is a clipped one and is superseded, the attendant will notice the fact the first time he looks in the hive, as the young queen will of course not be clipped. If none of the queens are clipped, it will frequently happen that a queen will be superseded without the knowledge of the bee-keeper.

Cutting Out Cells.—Some practice cutting out queen cells as a sole means of swarm prevention. At best this is an unsatisfactory plan. To be successful, every frame in every hive must be examined every eight days during the season. This entails so much work that it is almost entirely out of the question in a large apiary. An occasional cell will be overlooked and the bees will swarm in spite of the best attention.

If the bees have cast a natural swarm, one can then examine

the brood nest and cut out all queen cells but one. There will then be little danger of further swarming. (See Swarm Control under Comb Honey, Chapter IX.)

In small apiaries, operated as a side line, natural swarming will often prove to be the most desirable plan of increasing. If the bees are run for comb honey, the number of colonies are likely to double each favorable season, and sometimes there will be more than double the number of colonies at the close of the season that there were in the beginning. The extensive honey producer who makes bee-keeping a business, however, will wish to look for more certain methods of making increase.

Fig. 51.—Hiving swarm in straw skep in Europe.

Hiving the Swarms.—Hiving the swarms is usually a very simple matter. If the queens are clipped the hive from which the swarm issued may be removed, and the swarm allowed to return to the new hive set in its place as mentioned in a preceding paragraph.

If the queens are not clipped, the swarm will be likely to settle on a tree or on some other object near at hand. Small fruit trees about the apiary furnish the best clustering places, as the swarms can be taken down very readily (Figs. 5 and 52). If a comb containing brood is placed in the new hive, there is less danger that they will come out again and leave. Every bee-keeper of experience has lost swarms after thinking they were safely hived. Sometimes they will remain in the hive until the following day and then abscond. This is more frequently the case with after swarms.

Apparently as soon as a swarm is out, scouts go in search of a new location. It is well to hive the bees as soon as possible after they have clustered, and to move the hive to the place where it is expected to remain as soon as they are quietly settled, to avoid, if possible, the upsetting of the bee-keeper's plans by the return of enthusiastic scouts. At times a swarm will remain clustered for hours, and even over night, and be content when hived, while at other times they will leave with little ceremony within a few minutes. That scouts are searching for new quarters for days in advance of the issuance of the swarm, is evidenced by the fact that bees will be found in large numbers about an empty hive, or other available place for two or three days, when suddenly a large swarm will come in and take possession.

As soon as the cluster is formed, a sheet may be spread on the ground and the new hive set on it. The bees may be shaken on top of the frames or in front of the entrance (Figs. 53 and 54). As soon as a few bees go in they set up a joyful humming that attracts the others, and soon they will be moving in rapidly. If the queen gets inside all is likely to be well, but if she gets lost they will come tumbling out again within a few minutes. If swarms cluster in the top of tall trees, there is no way but to climb for them. They may be let down in a large basket with a rope tied to the handles.

The Alexander Plan.—The Alexander plan of making increase has come into general use in so many apiaries that no better plan, perhaps, can be offered.

When the colonies are nearly strong enough to swarm naturally, remove the colony to be divided from its stand, and put in its place a hive containing combs or frames of foundation. Remove the center comb from the new hive, and exchange it for a frame of brood from the old hive. Find the queen and put

Fig. 52.—Market basket swarm catcher.
Fig. 53.—Newly hived swarm.
Fig. 54.—Swarm caught in a sack, running into the hive.

her on this comb of brood in the new hive. Care should be used to see that no queen cells are left. On top of the new hive which contains the queen and the empty combs, place a queen excluder and set the old hive on top of it. After about five days look over the combs carefully, and if queen cells are started above the excluder the old hive should then be removed to a new location. If no cells are started the bees may be left until all young larvæ are capped, when they can be removed. At the end of twenty-four hours after removing the hive to the new location it should be provided with a queen or a ripe cell. Mr. Alexander preferred giving a laying queen, so that no time would be lost. He reported that with him this method entirely prevented swarming. His plan was to make the increase early in spring, as soon as the colonies were strong enough, but in many localities the divided colonies could not build up in time for the clover flow, and the crop would be short as a result. In such localities the division should be made toward the close of the main flow.

A Somewhat Similar Plan.—A very common practice in use for half a century is to take a single frame of brood from a strong colony and place this frame, together with the queen and frame of honey, in a new hive and add combs or frames of foundation to fill up the remaining space. The old hive is then removed to a new location, and the new hive placed on the original stand. The field bees will return to the queen in the new hive, on the old stand. This plan should only be undertaken in very warm weather, when there is less danger of loss of the hatching larvæ. The only difference between this plan and the one above described is that in this case the division is made at once instead of leaving the young bees over the new hive for a few days until the larvæ have been capped over. There is a greater loss of bees by this method than the former one, unless the operation is carefully performed, as there are not likely to be enough nurse bees left in the hive to care for the young larvæ. Divisions without provision for caring for all the young brood are expensive, and not to be recommended. If the colony is disturbed as little as possible in the operation, and the hive only opened to find and remove the queen and to take out the frame of brood that is exchanged for an empty comb, many of the bees will remain in the hive and there will be little if any loss of brood. The brood combs should be pushed together and the empty one placed at the outside of the hive, rather than to divide the brood nest. A queen should be provided for the new colony as soon as possible.

Divisions without Queens.—It is far more profitable to provide each new division with a queen, or at least a ripe queen cell as soon as possible. However, it often happens that some increase is desired when no queens or cells are available. If the bee-keeper will plan ahead, cells may easily be raised by the Miller method as described on page 122.

If one wishes to make a division without providing a queen, it may be done as follows: From your best colony take a frame of brood, being sure that eggs and newy hatched larvæ are present. Add empty combs or frames of foundation as in the other cases described, to fill up the space in a new hive. In the middle of the day, when bees are flying freely, remove your strongest colony without opening the hive or disturbing them more than is necessary some distance away, and place the new hive with a frame of brood, but no queen, where the strong colony stood. The field force from the latter colony returning to their old location will make the best of the situation, and proceed to rear a queen from the young larvæ, and by fall there will be a strong colony if conditions are favorable. The old colony which has been removed will lose their field force, and consequently will require some time to build up to normal condition again.

Another plan is to divide the brood from a strong colony into two parts, placing half of the frames in the old hive and the rest in a new hive. Both hives are filled up with empty combs or frames of foundation. No attention is paid to the queen, but care is used that eggs and very young larvæ are present in both hives. The two hives are then set closely together, side by side on the old stand, each occupying about half of the original space.

The field force will be about equally divided between the two hives. The one in which the old queen remains will build up much more rapidly than the colony that must rear a new one. While this plan may serve in an emergency, it is not to be recommended for general practice, as are none of the plans of increasing without additional queens, as too much time and energy is lost on the part of the colony in replacing the queen. If the honey crop is not to be considered, and increase alone is desired, then these plans might be permissible, but even for this purpose so much greater results can be obtained by rearing the queens in advance that it is the best practice.

Forming Nuclei.—When it is desirable to make increase in quantity, the best plan, perhaps, is to break up strong colonies for the purpose, and make as many divisions as possible without hope of their gathering any surplus honey. In this way rapid increase may be made with fairly satisfactory results.

To begin with, take the queen and one frame of brood from a strong colony, and add another frame of sealed brood from another colony, to give her enough bees with which to start housekeeping. Place her with the frames in a new hive in a new location, and shut the hive tight with grass to prevent her bees from returning to their old stand. If the weather is extremely hot there is danger of smothering, and in that case the bees can be placed in the cellar instead. At the end of two or three days if the bees have not gnawed out, the grass may be removed in the evening, after the bees have stopped flying (Fig. 55).

After the queen is removed the old hive should not be disturbed for several days. At the end of ten days the young brood will all be sealed, so there is a minimum of danger of loss of young bees. A number of ripe queen cells will also be present in the hive. If there are six frames of brood, it may be divided into three parts, placing two frames of brood in each hive. Care should be used to see that each hive contains at least one good queen cell. Empty frames may be added and the hives placed in the cellar or closed up tight with grass as before for two or three days. The entrances should be opened at night after the bees have stopped flying, to prevent a large part of the bees from returning to the old stand. If the bees first get out late in the evening, they will begin to carry out dead bees, and attend to other housekeeping duties, and by morning will have become accustomed to the new conditions, so that not nearly so many will return to the old stand as will be the case if they are released from confinement in the middle of the day. In this way four colonies should be secured from the one.

FIG. 55.—Nuclei in queen-rearing apiary.

It is seldom profitable to attempt such divisions of small or weak colonies, even in warm weather, as the amount of increase secured is too small to be profitable. Rather should the colony be left until it becomes strong before breaking it up. Rapid increase can be made from populous colonies during a honey flow, but when no honey is being stored it is difficult to get the nuclei to build up quickly, even though they be fed. One should always expect several weeks of some kind of honey flow after making increase.

In case of sudden check in nectar secretion for any reason,

the apiarist must be exceedingly careful or he will lose much of his newly made increase. Robbing is likely to be general when no honey is coming in, and for this reason all entrances should be contracted to about an inch in width or even less if the nuclei are very small. Combs of honey should be provided to all these weaklings to insure sufficient stores to enable them to continue brood rearing. Even then, if no honey is coming in the queen may stop laying and everything remain at a standstill until the flow again begins. In order to avoid this undesirable condition, it is well to feed a little sugar syrup each night after the bees have stopped flying. It is almost impossible to feed during the day when there is a dearth of nectar without starting robbing. If the dearth continues for a long period the apiarist may find it necessary to again unite his nuclei, and his labor will be for naught.

Supplying Empty Combs.—When making divisions by any of the above plans drawn combs should be supplied to fill out the empty space in the hives if possible. Weak colonies should not be taxed more than necessary in comb building. If drawn combs are not at hand, full sheets of foundation should always be used, for otherwise the comb built under such circumstances will be mostly composed of drone cells and of no value in the brood chamber. Drone comb is only valuable for storage purposes and can only be used in the extracting supers. Frames of drone comb in an apiary are always a nuisance, as the apiarist must constantly be careful lest they be slipped into a brood nest somewhere, or the queen going above into the extracting super shall make use of them. Fig. 91 shows two combs illustrating this point. The upper one is built on a full sheet of wired foundation and is composed entirely of worker cells. The lower one was built without foundation and is composed entirely of drone cells.

If, as frequently happens, the apiarist has made too many new colonies and they are not likely to reach the end of the season in good condition, he can take a frame or two of brood from each strong colony toward the end of the season and use them to strengthen the little colonies that have been building up from the nuclei formed from the earlier divisions. When the season of honey production is nearly over, a frame of sealed brood may be taken from a strong colony without injuring it in the least, as the bees will emerge too late for the honey flow and the colony will already be sufficiently strong to winter well. At the same time emerging brood will do wonders for the weak ones if given a short time before the honey flow ceases, and will be valuable at any time.

If a large amount of increase is made in one season it will be necessary to make a liberal allowance for expenses of queens, and foundation, and considerable feeding of honey or sugar are also likely to be necessary. Unless one has had considerable experience with bees, too rapid increase is likely to lead to disaster. For the average person the Alexander plan is perhaps the safest that can be recommended. It is better to undertake to make but two colonies from one at most, unless it is done by experts of long experience. If this division is made early and the two colonies become strong again while there is a considerable period of honey flow still to come, the same operation can be repeated a second time, thus giving four colonies in all from one to start with. There are important factors in making increase that are not readily apparent to the novice, even though he read directions carefully, and he should be content to go slow and advance surely rather than take the risk of closing the season with fewer colonies that he began it with.

Making Rapid Increase.—The following account of Dr. Miller's method of increasing from nine colonies to fifty-six in one season will show the possibilities of making rapid increase in a favorable season:

On June 12 the best queen in the apiary was taken from her hive and placed on a set of empty combs. Her brood was removed to the stand of another colony, which in turn was moved to a new location. There were thus three colonies instead of two.

The first had no brood, but the field bees would shortly return to make a broodless colony. No. 2 had no queen but would get the field force of No. 3, which had been moved to a new location and would require some time to recuperate. Seven colonies still remained which had not been touched. Each of these was examined, and wherever possible to spare a frame of brood it was taken away and given to No. 1, which had no brood. To begin with, he found only four frames, but this was given to the colony which had been robbed of its brood, being set on the top in a new hive body.

At the end of nine days a second visit was made. This time No. 2, which had brood but no queen, was divided into two parts, as by this time queen cells were present. The two nuclei were set in new locations and the brood and bees again taken from No. 1 and placed where No. 2 had been. The other seven colonies were again visited and such brood as they could spare was taken from them and given to No. 1. This plan was continued through the season, always leaving the queen at No. 1, so that the queen cells built on the combs in No. 2 were the offspring of the best queen. No. 2 did not at any time have any queen but was constantly building new cells and the other seven colonies were constantly (every nine days) drawn on for brood to replenish No. 1. In this way the colonies were at no time greatly weakened, excepting the nuclei made from No. 2. This is a very good plan of making rapid increase and at the same time a safe one, for if conditions suddenly become unfavorable the operator will not find himself with a large number of very weak colonies on hand, which must be united or fed.

INTRODUCING QUEENS

In making increase artificially by any plan an extra queen will be required to supply each new hive. If capped queen cells are given, the bees are likely to realize their queenless condition before the young queen emerges, so that she should be accepted without difficulty. This is a very common plan of pro-

INTRODUCING QUEENS

viding nuclei with queens, but several days' valuable time will be lost which might be saved to advantage by the use of laying queens if they are to be had.

It will be understood, of course, that the colony must be queenless or the introduced queen will quickly be killed. Apparently the bees recognize their queen by the odor common to all inhabitants of the same hive. A new queen lacking this odor will not be accepted. All methods of introduction depend for their success upon either leaving the queen with the colony long enough to acquire this odor before she is released, or creating some abnormal condition that will for the time being prevent the bees from recognizing the hive odor. The smoke method, water method, and several others come under this latter plan.

If an old queen is to be replaced, it is generally advised that she must be removed from the hive at least twelve hours before the new queen is introduced to give the bees time to miss her. Usually not less than twenty-four hours is allowed to elapse before requeening. Better results are likely to be obtained by requeening at once. If a queen cell is used the wait is desirable.

The novice will find it quite a task to locate the queen to be removed, but after a little practice it soon becomes an easy matter to find her. Gentle Italians usually remain quiet when the hive is opened and one can readily find her by looking first on one side and then on the other of the combs as they are removed from the hive. She will usually be found on a frame containing eggs and very young larvæ.

Black and hybrid bees that begin running from one side of the hive to the other as soon as it is opened, or boiling over the top as it is commonly expressed, will offer greater difficulties. It will sometimes be necessary to look the hive over from side to side several times before finding the queen. Professor Francis Jager recommends that the hive be opened very carefully and a little smoke driven in at the entrance. The bees will at once begin to boil over the tops of the frames and by looking there for the queen she can often be found without removing a single frame. This plan will not work well with Italians, for unless greatly disturbed they do not run about much, but simply dive into the cells and begin to take up a load of honey.

The Cage Method.—By far the most common method of introduction is the cage method, and it is generally regarded as the

Fig. 56. Miller Queen Cage.

safest method as well. By this method the queen is confined in a cage (Figs. 56 and 57) which may be placed between the combs in the hive for two or three days before she is released. If she comes by mail in the ordinary mailing cage there will be a quantity of candy between the queen and the opening which is closed by a cork. If the colony has not been queenless the cork

Fig. 57.—Benton queen cage. This is the cage usually used for sending queens by mail

may be removed at once unless the candy is nearly eaten through, in which case the cork had best be left in for a day or two. Usually it will require two or three days for the bees to eat away the candy and to release her. In the meantime she will have acquired the common hive odor and the bees will have become familiar with her, so that there is little danger but that she will be accepted.

Occasionally there is a colony, after being for some time queenless or when there is a dearth of honey and the bees are not in good temper, which will destroy the queen by whatever method she be introduced. After a failure or two, one hesitates to risk other valuable queens, and it seems advisable to unite the bees with some other colony rather than to bother further with them.

For use in the apiary where the precautions necessary in sending bees by mail are not necessary, the Miller cage is commonly used (Fig. 56). This cage has a larger opening. After the queen is caught and placed in the cage she is placed in any colony, simply laying the cage on top of the frames. Since the bees cannot destroy her she will be safe, for strange to say they will feed her through the meshes of the wire. When wanted to requeen a colony or to give to a nucleus, the cage is placed in the colony where she is expected to remain, until such time as the bee-keeper thinks best to release her. The cork is then removed and she may be allowed to escape. It is a common plan to fill the opening with honey comb which will require a few minutes to remove, thus giving time for the bees to become quiet again after the hive is closed, before she comes out.

The queen newly introduced is likely to be a victim of any excitement in the hive, and experienced bee-keepers usually are careful not to open the hives for several hours or better yet, for a day or two after a queen has been introduced so that she may become fully accustomed to the new conditions before being disturbed. It is easy to ascertain whether she has been accepted by examining the space in front of the hive. If she has been killed she will be found on the ground in front of the hive.

Miller Smoke Method.—During the past year there has been much discussion of the value of this method. Some bee-keepers report great success and feel that it is the ideal method. Others report failure. Difference in conditions will account somewhat for the difference in results, as any plan will work much better in the hands of expert bee-keepers and under favorable conditions than otherwise.

Mr. Miller describes his method in "Gleanings in Bee Culture" as follows:

> A colony to receive a queen has the entrance reduced to about a square inch with whatever is convenient, as grass, weeds, rags, or wood, and then about three puffs of thick white smoke—because such smoke is safe—is blown in and the entrance closed. It should be explained that there is a seven-eighth-inch space below the frames, so that the smoke blown in at the entrance readily spreads and penetrates to all parts of the hive. In from fifteen to twenty seconds the colony will be roaring. The small space at the entrance is now opened; the queen is run in, followed by a gentle puff of smoke and the space again closed and left closed for about ten minutes, when it is reopened and the bees allowed to ventilate and quiet down. The full entrance is not given for an hour or more or even until the next day.
>
> The queen may be picked from a comb and put in at the entrance with one's fingers, or run in from a cage just taken from the mails, her attendants running along too. The result is the same.

If directions are followed explicitly Mr. Miller claims that results will be as good or better than with any other plan. The author has given the plan a limited trial with good success, but it was during a honey flow when conditions were so favorable that there was little difficulty in introducing by any method. So many failures have been reported by experienced bee-keepers that the novice is cautioned against placing too much confidence in it to begin with. It will be safer for him to follow the directions on the cage in which his queens are received until he has had considerable general experience.

During a good honey flow when the bees are storing heavily there is little trouble in introducing queens by any method. On one occasion the author returned home after an extended absence and with but a few hours time introduced seventeen queens, many of which were given to strong colonies to replace the queens already in the hives. As it was necessary to leave again shortly, time was an object and there was no opportunity to leave the colonies queenless for even a few hours. The hives were opened and as fast as the queens were found they were removed and the new queens run in between the frames and the hives closed again. The bees had been given a little smoke to quiet them when the hives were opened and the colonies were somewhat disorganized by the removal of the frames to find the old queens, but in every case the new queen was accepted without accident. A few were run in at the entrance after the hive had been closed again and were followed by a puff of smoke, but the entrances were not closed or even contracted, yet the results were as above stated. Such results could not be obtained except under the most favorable conditions. The bees were simply too busy to attend to anything but the storing of honey and the new queens attracted no attention apparently.

The value of a method can never be demonstrated until it has been tried under many and varying conditions and especially under adverse conditions. The smoke method is, as yet, not fully vindicated, under general conditions.

Water Method.—*The American Bee Journal* for March, 1915, contained the following method as practised by the Southwestern Bee Co.:

> The procedure is as follows: Kill the old queen; remove all frames from the hive and shake into the bottom of the box with a sharp jar all the bees possible. Sprinkle the mass of bees on the hive floor with water until they are *soaking wet*. The secret of success is in the use of plenty of water; there is no danger of overdoing this part. Wet the new queen thoroughly and put her on the pile of wet bees. Put back the combs into the hive and the job is finished. We have been using this method for several seasons and have found the method successful with virgins, with laying queens, and with queens received in cages by mail.
>
> When honey is coming in, any time of the day will do for the work of introducing, but in times of dearth it is better to wait until about an hour before dark.
>
> The chief value of this method is that there is no time whatever lost and the new queen is immediately accepted and ready to go to work.

Honey Method.—Another plan based somewhat on the same general principle is to drop the queen into a little dish of honey, pushing her clear under. She will be so messed up that the bees will immediately begin cleaning her up when she is placed in the hive, apparently never dreaming that she is an alien. This is anything but an attractive way to treat a valuable queen but good reports are given by those who have tried it. It is hardly likely to become popular because of the fact that most bee-keepers will hesitate to subject a queen to such treatment.

Flour Method.—Dropping the queens into flour has also been tried with success in some cases. A combination of the flour and smoke methods is also practised successfully at times. By this plan a colony is smoked in the manner described previously, and just before running the queen into the hive she is dropped into a dish of flour. None of these methods seems to give the satisfaction under all conditions that the cage method does.

Sure Plan for Valuable Queens.—When one buys a queen of more than ordinary value and is willing to go to some extra trouble to insure her safety, there is an old plan of taking from strong colonies two or more combs of sealed brood and shaking off every live bee. There must be no unsealed brood or it will die from lack of nursing. These frames of sealed brood are placed in a hive body without a bottom board and this body is placed over another colony with ordinary wire screening placed over the top to prevent the bees from making trouble, while at the same time furnishing plenty of heat to the brood above the wire cloth. The queen is placed on top of one of the frames, or her cage is opened and she is permitted to run between the frames and the cover placed on the hive. The young bees as they emerge will be confined to the upper hive body with the new queen and she will shortly be surrounded with a nice little cluster. If desired, after about five days the queen, having been removed from the colony below, the wire cloth may be removed and two or three days later after the colony has become accustomed to the new queen the bees can all be shaken from the upper combs and the upper hive body removed. If desired to make a new colony, the upper hive body can be removed to a new stand and additional frames of brood given to strengthen them.

This is about the only plan that is regarded as entirely sure under all conditions. In this case there are no old bees in the hive when the queen is introduced and the emerging brood will make no trouble.

Fundamentals of Successful Introduction.—The following extracts from an editorial by the late W. Z. Hutchinson in the *Bee-Keeper's Review* in 1891 cover the ground very fully:

> To introduce a queen to a colony of bees, two things must be well considered—the condition of the bees and the condition of the queen. The condition and behavior of the queen is very important. If the queen will only walk about upon the combs in a quiet and queenly manner, and go on with her egg laying, she is almost certain to be accepted if the other conditions are favorable. Let her run and "squeal" (utter that sharp "zeep, zeep, zeep") and the bees immediately start in pursuit. Soon the queen is a mass of tightly clinging bees and the only course is to smoke the bees severely until they release the queen from their embrace, when she must be re-caged for another trial.
>
> So far as the queen is concerned, it is important that she be brought before the bees in a natural manner, in such a way and in such a place as they would expect to meet her. When clipping queens I have replaced them by dropping them upon the top bars, or at the entrance of the hive, when the bees would immediately pounce upon them as intruders. A puff of smoke would cause the bees to "let up" when the queen would walk majestically down upon the combs or into the hive, as the case might be, and here she would not be molested, because the bees here found her where they *expected* to find their queen.
>
> No definite length of time can be given as to how long a queen should be caged before she is released. The behavior of the bees is the best guide. If they are "balling" the cage, clinging to it in masses, like so many burdocks, their behavior indicates what the queen would have to endure were she within their reach. The operator must wait until the bees are in a different mood; until they are walking quietly about the cage, as unconcernedly as upon the combs of honey—perhaps the bees may be offering food to the queen and caressing her with their antennæ. This shows that the bees are favorably inclined toward the queen and that it is safe to release her.
>
> To be successful in introducing queens that have come from a distance, the condition of the colony must be well looked after. It is better that they should be *hopelessly* queenless. Let it build a batch of queen cells, and remove them after the larvæ are too old to be developed into queens; then the bees are almost certain to accept a queen if given to them in proper manner. I had sooner release a queen after the bees had discovered the loss of their old queen, and before they had begun the construction of queen cells, than to release her after the cells were under way, *unless* I waited until the cells were sealed over and had been removed.
>
> Bees are in a much more amiable mood when honey is coming in freely. Don't attempt to introduce queens when no honey is being gathered, without feeding the bees two or three days before the queen is released.

So much has been written about the introduction of queens and so many plans are in use that it is difficult to give a comprehensive review of them without leaving some confusion in the mind of the reader. If, as is commonly believed, the hive odor is the means by which the bees recognize the members of a common community, the great object to be attained by any method is that the new queen shall acquire this peculiar odor as quickly as possible. More than twenty years ago it was recommended that to assist in accomplishing this result the queen to be removed from the colony to be requeened should be confined for a time in a cage. She is then removed and the new queen placed in this same cage by means of which she is introduced to the colony. This method has been reported as very successful by bee-keepers for many years past. This is essentially the ordinary cage method with the exception that the former queen is confined in the same cage in which her successor is to be introduced for a time before she is destroyed and the new queen placed therein.

When a queen is to be introduced by any of the direct methods it will be much help if she is confined by herself for at least thirty minutes without food. Being hungry she will at once solicit food when she comes in contact with the workers and will much more likely be accepted.

QUEEN REARING

Although commercial queen rearing is a business by itself that would require a volume for exhaustive treatment, the bee-keeper's education is not quite complete until he has learned to rear his own queens, even though it may not be advisable for him to do so to any extent. Most productive bee-keepers feel that they can ill afford the time for extensive queen rearing at the busy time of year when they can best be reared, and prefer to buy them from some regular breeder. There are times, however, when one can rear his own queens to advantage, and it is always well to be prepared to supply a limited number for special purposes or to meet emergencies.

Some of the most successful honey producers feel that only by breeding from selected stock which has been tested under their own conditions can they secure best results, and for this reason alone are willing to rear their own queens. The difficulty of controlling the male parentage makes the breeding of bees much harder than the breeding of poultry or farm animals which are under the absolute control of the farmer.

The fact that the characteristics of the male offspring of a queen are controlled rather by the mating of her mother than by her own mating adds to the difficulties. It is a well-known fact that an unfertilized queen will produce drones and this leads to the belief, now generally accepted, that her mating does not directly affect the eggs from which the drones are hatched. Consequently they are only influenced through her female offspring, and results are only apparent in the following generation.

That progress is being made in the improvement of the honey-bee there can be no doubt, and by breeding only from the best queens something is sure to be accomplished even though the male parentage be uncontrolled.

Scientific men are giving a good deal of attention to the problems presented in breeding bees, and it is only a question of time until methods suited to the conditions to be met will be devised and scientific bee-breeding will be an accomplished fact.

About all that the bee-keeper can do is to see that all his colonies have queens of good stock, so that drones from worthless stock will not be present, and breed only from his best queens. Even then the queen may fly some distance from home on her mating trip and meet a black drone from some neighboring apiary. Some bee-keepers make it a practice to see that all small apiaries within two or three miles are requeened with good stock, even furnishing the stock when necessary and doing the work to save the annoyance of mismated queens. In some localities this would be easy of accomplishment, while in others where large numbers of hives are present it would be a big undertaking.

Necessary Conditions for Rearing Good Queens.—The best queens are reared under the swarming impulse, or in other words under natural conditions. When the bee-keeper would resort to artificial conditions to rear queens for the purpose of increasing his stock or replacing inferior queens, he must make conditions as favorable as possible.

The weather should not be cold or unfavorable when queen rearing is undertaken, the best queens only should be used as mothers, and the cells should be built in strong colonies. One would hardly expect to get best results unless the bees were storing some honey also.

Miller Plan.—What is known as the Miller plan, or some modification of it, is perhaps the best method for ordinary non-commercial purposes. The best time for requeening is perhaps about ten days before the close of a honey flow. Checking the egg laying of the queen at this time will have no influence on the size of the crop, as the young bees hatched from eggs laid after this time would appear after the close of the harvest. The bee-keeper will begin to make plans for requeening then about three weeks before the expected close of the flow.

From the center of the brood nest of the colony containing the best queen the bee-keeper removes a frame of brood and replaces it with a partly drawn comb or a half sheet of foundation or even a frame with starters. This will quickly be utilized, and if foundation is used the bees will draw it out and the queen will fill the cells with eggs. Old combs should never be used for this purpose, as they do not furnish suitable conditions for building good cells. The author prefers a partly drawn comb and, lacking that, uses foundation. In a few days this comb will be filled with eggs and hatching larvæ. The next move is to remove the queen from some strong colony and take away a frame of brood from the center of the brood nest and replace it with this partly drawn comb filled with eggs from the best queen. The colony finding itself queenless will at once start queen cells along the edges of this new and tender comb which furnishes ideal conditions for cell building. At the end of ten days the cells will be nearly ready to hatch and should be removed to avoid the danger of swarming or of an emerging queen destroying the others.

Queens to be replaced should have been removed the day before so that the bees will have missed them and be ready to accept the cells. The cells should be cut from the comb, care being used not to cut too close and injure the young queen. They may be fastened to the side of a brood frame with a toothpick or simply dropped between two frames above the center of the brood nest.

If they are to be used for making increase, the colonies to be broken up into nuclei should be previously made ready. The advantage of using the ripe cells is that much less trouble is necessary than when mating the queens in baby nuclei and then later transferring them to the full colonies to be requeened. If used in nuclei for purposes of increase they emerge, and are mated from the hive in which they will remain, thus saving the trouble of introduction. It is much easier to get a cell accepted than a virgin queen or even a laying queen.

Dr. Miller recommends the use of a colony that has started queen cells in preparation for swarming as cell builders when available, as conditions in such a colony will approximate the natural method of queen rearing.

Commercial Methods.—In order to succeed commercially the queen breeder must be able to supply queens in considerable numbers with regularity throughout the season. If the honey flow is checked he must feed his colonies freely in order to continue to supply his orders. While the foregoing plan is well suited to rearing a few dozen queens to supply the needs of one's own apiary, it would be entirely too slow for commercial purposes. Then the man who rears queens for sale must be prepared to get them mated safely, which requires additional equipment in the way of nuclei, etc. In making nuclei for mating purposes one can get a much larger number from breaking up a single colony of bees than would be possible where they were used for increase, each of which was expected to build up to a full sized colony. Often the bee-keeper will not use more than a dipperful of bees for a nucleus used for mating purposes only. It is a common practice to shake the bees from a single frame to use for this purpose and to unite them with full colonies again in the fall.

There are two plans in common use among commercial queen breeders. One is known as the Alley plan and the other as the Doolittle plan, after the men who originated them. While in many cases both plans have been greatly modified since first made public, the general principle remains.

The Alley Plan.—If one wishes to make use of this plan the first thing is to remove a brood comb containing eggs from the brood nest of the colony led by the best queen. No bees should be retained on the comb, as it will be necessary to cut it in strips. Each strip contains just one row of cells. With a sharp knife cut through the row above and below, saving every other row for use. This cutting must be done very carefully to avoid injury to the delicate comb and the eggs it contains. After the strips have been cut they are laid down and the cells on one side cut down to about a quarter of an inch of the foundation or center of the comb. With a match destroy every other egg in this side. These shallow cells can now be readily built into queen cells by the bees as shown in Fig. 58. This same picture shows how the strips are fastened to the bottom of a shallow comb, so that they will be in the center of the brood nest of the hive in which they are placed. If the knife is kept hot there is less danger of jamming the cells when doing the cutting. The strips can be fastened by means of melted beeswax, which will adhere to the wood strip on the lower side of the comb. Mr. Alley fastened them directly to the comb without any wood strip. The combs with strips are then given to queenless colonies.

While most commercial establishments use some modification of the Doolittle cell cup method, the Alley plan is still used by some queen breeders who prefer it to the other.

The Doolittle Method.—A great advance has been made in

THE DOOLITTLE METHOD

queen rearing since Doolittle hit on the plan of making artificial cell cups. His plan as first used was to take a small stick with round end about the size of the bottom of the cell and after dipping it in water dip it in melted wax. Several times it was thus dipped, each time not quite so deep as the time before, thus leaving the base much thicker. When it was of the required thickness it was removed and others made in similar manner. These were fastened in frames which would fit into the hive in place

Fig. 58.—Queen cells by the Alley plan.

of a regular brood frame. A newly hatched worker larva, together with a small amount of royal jelly, was placed in each cell, with the result that very good queens were reared.

Wood cell cups and artificial wax cells are now offered for sale by dealers at prices that will no longer justify the beekeeper to make his own cells. The cell cups are listed in dealers' catalogues at about $2.00 per 1000, which is cheaper than the average person can make them at home. The cups are pressed into the wood cell holders by means of a wood plunger and are ready for grafting, as the placing of the larvæ is called.

The elaborate descriptions of this system are disconcerting to the novice but in reality it is quite simple. It is rather a delicate task to transfer the newly hatched larvæ from the worker cell to the artificial cell in the wood cup, but a little experience will make it easy. By the present method there are no cells to dip or other complicated processes to confuse the inexperienced. He needs only to place as many of the wax cells in the wood cups as he wishes to use. He then places these in the frames in which they hang in the hive and transfers a baby bee and a drop of royal jelly to each one.

Care of Cells.—The real problem by this method is to get colonies in proper condition to care for the large numbers of cells which the commercial queen breeder must constantly have in order to get a sufficient number of queens to make it profitable. While some queen breeders rear their queens in small nuclei, it is the general opinion that the best queens are reared in strong colonies.

A colony can be made queenless and after twenty-four hours be given a frame of these prepared cells. They are likely to be accepted and cared for and a second lot can be given when these are taken away. However, the leading queen breeders have been seeking a method of safely finishing these cells in strong colonies with laying queens so that the queens will be reared under similar conditions to those reared when the bees are preparing to swarm. The Roots, who are extensive queen breeders, practise making two-story colonies with the queen in the lower story and an excluder between the two. The brood is raised into the upper story so that the queen will go on laying below but no new brood will appear above. The frame with prepared cells is placed in the center of the brood above the excluder and the bees finish the cells nicely. When one batch is removed another is given in place of it, and when all the brood is hatched above, the brood from below is again lifted to insure proper attention to the queen cells. It seems to be necessary to keep brood above to secure good results.

Some queen breeders secure similar results on one side of the hive by using a queen-excluding division board to prevent the queen from reaching and destroying the cells.

QUESTIONS

1. When will natural swarming be most apparent?
2. What conditions favor swarming and what conditions act as a check?
3. Mention the advantages of clipping queens.
4. Why is cutting out queen cells an unsatisfactory method of swarm control?
5. Describe the common method of hiving swarms.
6. Outline the Alexander plan of swarm control.
7. Note the differences in a similar plan long in general use.
8. How are divisions made without queens?
9. Discuss the formation of nuclei.
10. Describe Miller's plan for making rapid increase.
11. Discuss the different methods of queen introduction.
12. What conditions are necessary for rearing good queens?
13. Discuss the different methods of queen rearing and give the advantages of each.

CHAPTER VIII

FEEDING

Probably one-third of the total annual loss of bees is the direct result of carelessness on the part of the owners in failing to provide stores at the proper time. In the spring such large quantities of honey are consumed in early brood rearing that a few days of unfavorable weather will bring a colony with a small reserve supply to the verge of starvation. Thousands of colonies are lost from this cause alone. Then it is nearly always the case that some colonies will go into winter quarters with an insufficient food supply, unless fed, and will die for lack of stores before spring.

When making increase or rearing queens a check in the honey flow will make it necessary to continue to feed the colonies in order to maintain normal conditions and get best results.

From the above statement it will be seen that feeding at the proper time is a matter of the greatest importance. Perhaps more needless loss is caused by a lack of appreciation of this fact on the part of the average bee-keeper than any other.

Good Honey the Best Feed.—As mentioned incidentally elsewhere, the author regards good combs of sealed honey as the best feed for all times excepting when it is desired to feed slowly to stimulate brood rearing. The far-sighted bee-keeper will retain a supply of extracting combs filled with sealed honey for this purpose. They are always ready and can be placed where needed with but a moment's time.

There are localities where the bees gather honey-dew and honey of low grade that gives unsatisfactory results in wintering, where it is sometimes considered advisable to extract the honey and feed sugar syrup. Such places, fortunately, are not many. The storing of syrup is quite a tax on the bees and cannot but result in a decrease in the number present by wearing them out prematurely.

Preparing the Syrup.—When honey is not to be had syrup made from granulated sugar is the best substitute. Molasses or other cheap syrup should never be used, as such substitutes contain wastes that are bad for the bees and in the end are no cheaper. Practical bee-keepers are now agreed that if a substitute for honey must be used, granulated sugar is not only the best and cheapest thing but about the only safe feed commonly available.

If the feeding is done for the purpose of providing a reserve food supply, as in winter preparations, it is considered best to use a thick syrup composed of about equal parts sugar and water. Some use a syrup as thick as two parts sugar to one part water.

According to C. E. Bartholomew, a syrup made of 85 parts sugar by weight to 50 parts water will neither granulate nor ferment by standing. If it is desired to make a large quantity in advance to keep on hand for use as needed, this proportion should be used.

The syrup can be prepared by dissolving the sugar in hot water in the proportions desired and hastening the process by boiling, or cold water can be used, stirring the syrup from time to time until the sugar is completely dissolved. If a thicker syrup than equal parts of sugar and water is desired, it is best to use hot water, as it is difficult to dissolve larger quantities of sugar in cold water.

Care should be used that the sugar is not allowed to burn, as burned sugar is injurious if not fatal to the bees.

For feeding colonies that are rearing queens, or building up nuclei when a food supply is present in the hive but stimulation is desired, a thin syrup is usually used. It should not be thicker than one part water to one of sugar, and even thinner syrup is often used for this purpose.

Feeding for Reserve Supply.—When it is desired to feed colonies that are short of stores either for wintering or for spring brood rearing, it is desirable to feed as fast as they are able to take care of the syrup. The quicker the job is finished the less the colony will be disturbed and excited as a result. For this purpose some of the feeders holding a large quantity are best. For outdoor wintering in the north it is usually estimated that at least twenty-five pounds of honey will be required, and from fifteen to twenty in the cellar. It is much safer to have from thirty-five to forty pounds of stores for outdoor wintering and at least twenty-five for the cellar. In the South where the bees are active all winter, even larger quantities will be consumed.

The bees should not be provided with such a quantity of stores that there is no clustering space under the food supply in the center of the hive. The winter nest, as these vacant cells are called, permits the bees to conserve the heat by close contact. If they are compelled to cluster between full combs of honey the heat will not be sufficient to warm the cold mass between the small bunches of bees on opposite sides of the combs. However, if feeding is done early the bees will arrange matters nicely and remove sufficient honey from the center of the hive to form a clustering place.

Feeding to Stimulate Brood Rearing.—Some bee-keepers advocate the feeding of colonies with a plentiful food supply early in spring to start brood rearing. This is likely to be a mistake. If there is an abundance of feed in the hive a good queen will usually begin laying as fast as weather conditions will permit. In case a colony is too slow to begin operations, the hive may be opened and the cappings cut from part of one comb. The bees will feed the queen more liberally from this uncapped honey and she will lay more eggs as a result.

When the honey flow is checked after nuclei have been formed, it is desirable to keep the queens laying as fast as possible in order to insure that the colony will be strong enough to winter. The same thing applies when queen cells are being built—the bees must continue normal activities. For this purpose some form of feeder that will supply a small quantity continuously is best. With a small amount of syrup coming in the bees will continue as though honey was being brought to the hive.

FEEDERS

Great care must be used in feeding weak colonies or nuclei during a dearth of nectar, as robbing is started very easily at that time and the little colonies may be easily lost as a result. At

Fig. 59.—The Minnesota bottom feeder. This feeder is safe from robbers and holds a sufficient quantity at one filling.

times robbers are so persistent in sneaking about every crack that it is unsafe to open the hives excepting just at nightfall. Consequently the feeders should be filled at that time.

FEEDERS

For feeding large quantities of honey for reserve supply there are two good feeders on the market. Either will hold as much at a single filling as is likely to be fed to one colony. One of these is placed in a super on top of the hive and the other is set under the hive in place of a bottom board.

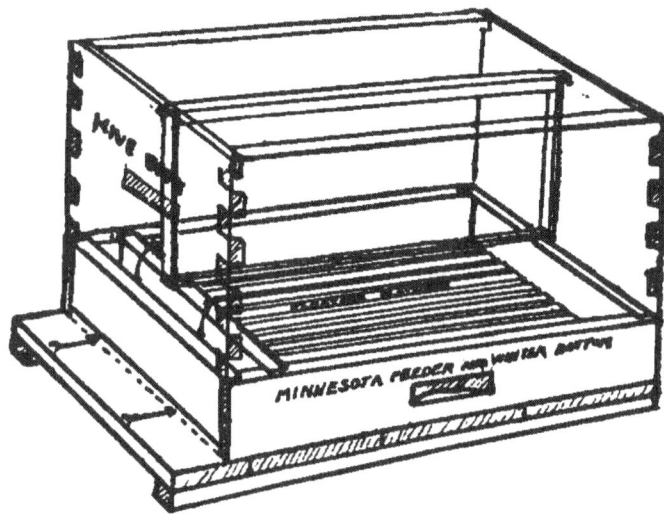

Fig. 60.—The Miller feeder is set in a super on top of the hive. This is one of the best feeders.

Minnesota Feeder.—This feeder, invented by L. D. Leonard, has some advantages over any other on the market. Fig. 59 shows how it works. By referring to the illustration it will be seen that a large box, with slatted float through which the bees take up the feed, occupies the entire bottom surface of the hive, excepting a small space for entrance. This feeder has a capacity of about thirty-five pounds, which will be as much as a colony can use to advantage under any circumstances, at one time. The syrup is directly under the cluster and can be taken up when the weather is too cold for them to break the cluster. The entrance opens outside in the usual way but leads upward instead

FIG. 61.—Tin pan feeder in super.

of backward. This bottom feeder can be left in place all winter, or all year if desired. The space below gives the bees a clustering place below the frames, which is a decided help in swarm prevention and also of value in wintering. In winter the wind will not blow directly into the hive, which is an advantage.

After the feeder has been filled and the hive placed in position the feed cannot be reached from the entrance, directly, which is of much help in preventing robbing.

Miller Feeder.—Fig. 60 shows the construction of the Miller feeder. There are two compartments on either side, each holding ten pounds or more of syrup. In the center is a passageway for the bees to reach the syrup going directly above the cluster. In

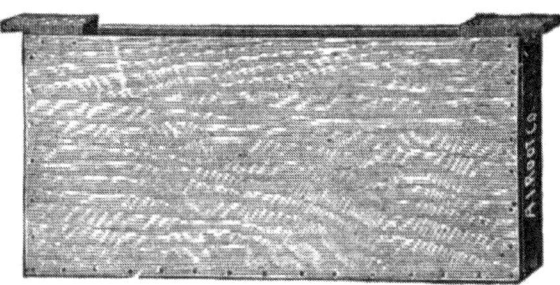

Fig. 62.—With the addition of a float to prevent drowning of the bees the Doolittle division board feeder is fine where small quantities of feed are to be given.

this way the warm air rising above the brood nest makes it possible for them to reach the food when the weather is quite cool. Twenty or more pounds can be fed at one filling if desired.

Tin Pan Feeder.—One of the most common ways of feeding small quantities of syrup is to use a tin pan in an empty super.

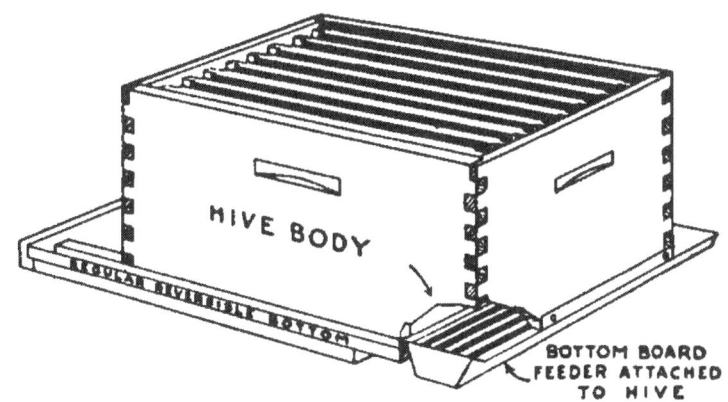

Fig. 63.—Metal feeder after the Alexander idea.

If the weather is warm the super can be placed on top of the hive and if cold underneath. Over the pan of syrup is spread a thin cotton cloth with edges hanging down all around to make it easy for the bees to get into it. They suck the syrup through the cloth without danger of drowning (Fig. 61).

Doolittle Division Board Feeder.—This is quite a popular feeder for colonies that are to receive but a small supply. As shown by Fig. 62 it takes the place of a brood frame in the hive. After it is filled the cover can be replaced and the colony left in the same snug shape as though no feeding was being done. It is nothing more nor less than a tight box of the size and shape of a brood frame. A float should be used to prevent drowning the bees.

Alexander Feeder.—The Alexander (Fig. 63) is fine for feeding nuclei in large numbers. Fig. 64 shows how it is fastened

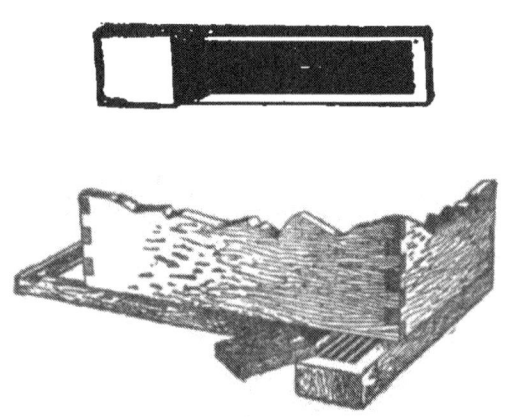

Fig. 64.—The Alexander wood feeder is good for stimulative feeding for rearing queens or making increase.

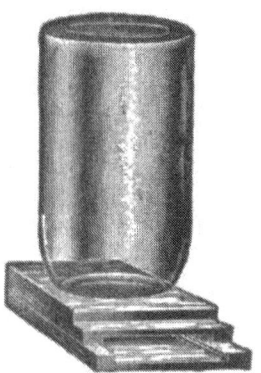

Fig. 65.—With this entrance feeder one can see at a glance how much feed remains to be taken. There is greater danger of robbing in using entrance feeders than the others.

to the hive by pushing the bottom board forward and putting it on immediately behind it. In this way the feed is away at the back of the hive safe from robbers and as the feeder opens on the outside it can be filled without opening the hive. With a large pail of syrup and a dipper, or a tea kettle, each of these feeders can be supplied in a moment's time. The one shown in the figure attached to the hive is made of metal. The tops are wider than the bottoms so that they may be nested together for convenience when not in use. The other illustration shows the same feeder made of wood which is most commonly used.

Entrance Feeder.—This feeder, commonly called the Boardman feeder, utilizes a common fruit jar as a container for the liquid (Fig. 65). Small holes in the screw top permit the feed to drip out slowly. The wood projection slips into the entrance so that the bees can reach the feeder from the inside of the hive while guarding the entrance from robbers. Glass fruit jars are common utensils in every household, so that all that is necessary to buy is the wood block and the special cap to fit the jar. One can see at a glance just how much feed still remains and by plugging up part of the holes it can be made to feed as slowly as desired. This is a popular feeder for making increase.

QUESTIONS.

1. When is it necessary to feed?
2. What is the best feed for bees?
3. How is syrup prepared for feeding?
4. Describe different methods of feeding for reserve supply and for stimulation.
5. Discuss the different feeders in common use.

CHAPTER IX

PRODUCTION OF COMB HONEY.

The successful production of comb honey requires more skill, perhaps, than any other branch of agricultural pursuit. Under certain favorable conditions it is a very easy matter and anyone who will supply sufficient supers will get a good crop. Such conditions, however, are of rare occurrence and the average season in the average locality gives abundant opportunity to develop the resources of the producer to the utmost.

The man who specializes in comb honey will usually produce small quantities of extracted honey also, while the extracted honey producer will have no occasion to produce sections, unless he especially wishes to have some of both.

As to whether one should specialize in comb or extracted honey will depend upon many things. The skill of the operator is an important consideration, as extracted honey does not require as careful attention to details as the production of comb honey. The amount of the crop, source from which it comes, and the market which is available, all should be taken into consideration. Comb honey as a rule commands a more ready sale and does not require the expensive machinery necessary to satisfactory production of the extracted product.

One of the most important factors is the nature of the honey flow. If one lives in a region where the flows are long and very light, it is difficult to get well-finished sections, and extracted honey will nearly always prove more profitable. If, on the other hand, the flows are short and very rapid, so that honey is piled up so fast as to make the bees fairly dizzy with the excitement of it, sections will be nicely finished and a large part of the crop can be made to grade fancy or number one. Under circumstances of this kind sufficient wax is secreted to build the combs with little noticeable tax on the production, and comb honey will probably be more profitable.

Rapid flows like those that sometimes come from basswood, when a single strong colony will store from ten to twenty pounds daily for a week or two, are the delight of the heart of the comb honey producer.

Market Demands.—Most markets favor light colored honeys, usually called white, which are of a mild flavor. As a rule dark and strong honey will sell more readily to buyers of extracted honey than in the comb. Where the market demand is for dark honey, as in some buckwheat sections, this will make little difference.

It is a common thing to find an established bee-keeper changing from the production of one to the other to supply a ready market. If all these questions are carefully studied in the beginning, much unnecessary expense will be saved.

If but a few colonies are to be kept to supply the family table, comb honey is to be preferred under almost any ordinary circumstances. Section honey is more attractive to most people, and less expensive equipment will be required. The fact that the sections are not always well finished will be of little matter for home use, although very vital in marketing.

EQUIPMENT FOR COMB HONEY PRODUCTION

The question of a hive is touched upon incidentally in the chapter on starting with bees. There is something to be said in favor of using the particular kind of hive in general use in the locality in which one lives. With hives of a pattern uniform with those in general use bees can be sold for better prices and one can make use of bees which he may chance to buy to much better advantage. However, one can ill afford to use a poor hive simply because it is in general use, as the best equipment makes possible easier manipulation and better crops.

For a time there was quite a tendency to adopt a hive of small size for the production of comb honey. The eight-frame Langstroth and the Danzenbaker hives were very popular and many bee-keepers adopted them only to discard them later. The principal argument in their favor is to the effect that the queen will require most of the available space in the brood nest and that the bees will quickly be forced to begin storing in the supers. Thousands of colonies of bees have died as a result of the adoption of this hive by persons who were not fully prepared to give proper attention to their bees. Nearly every year a part of the colonies in any apiary will not leave a sufficient amount of honey in the brood chamber of these small hives to winter on, and unless fed will die as a matter of course from lack of food.

The tendency to swarm is much greater in these small hives than in larger ones, and swarm control is important to the comb honey producer. Most authorities now agree that the ten-frame Langstroth hive is better for all purposes than a smaller one. The reason the Langstroth is recommended in preference to others of the same size is because its use is so much more general than any other hive (Fig. 66).

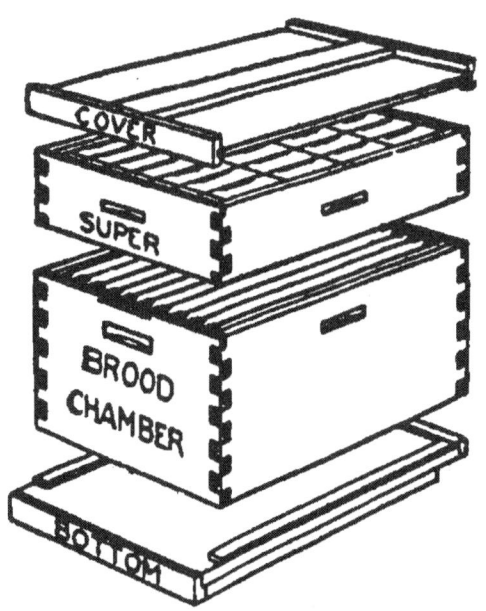

FIG. 66.—Parts of a comb honey hive.

If the small hive is used two hive bodies instead of one should usually be used for wintering, when packed outside.

It may be said in passing, however, that C. C. Miller, who has produced larger average yields of comb honey than any others on record, uses the eight-frame Langstroth hive. It is doubtful whether he would use such a small hive if he was starting again. While the hive is important, the management after all is the determining factor in measuring the profit of an apiary, next to the available supply of nectar in the field.

Sections.—Next to the kind of hive the question of the kind

SECTIONS

of supers and sections to adopt must be considered. This is a matter that must be determined by individual preference, for there is no one particular best section. The bee-way section is perhaps more widely used than any other, although the plain section would be a close second.

The bee-way sections are made in four styles. The kind in most general use is cut with bee-ways in two of the four sides

FIG. 67.—Strong colonies for comb honey production.

(Figs. 68 and 69). This gives the bees access to the sections from below and permits them to pass through to the super above. A few with only one bee-way are used. These permit the bees to reach the super but no passage-way is provided for them to go above past each section. The four bee-way sections permit the bees to pass from one section to another in the same super without going above or below.

140 PRODUCTION OF COMB HONEY

Fig. 68.—Comb honey supers.
Fig. 69.—Comb honey super dissected.

SEPARATORS 141

In general the two bee-way section 4¼ inches square and 1⅞ inches wide is to be recommended to those wishing to adopt the bee-way section. This is the section usually sent out by a factory receiving an order when there is no stipulation as to the kind wanted. This indicates their general popularity.

The 4 × 5 plain section of 1⅜-inch width seems to be the

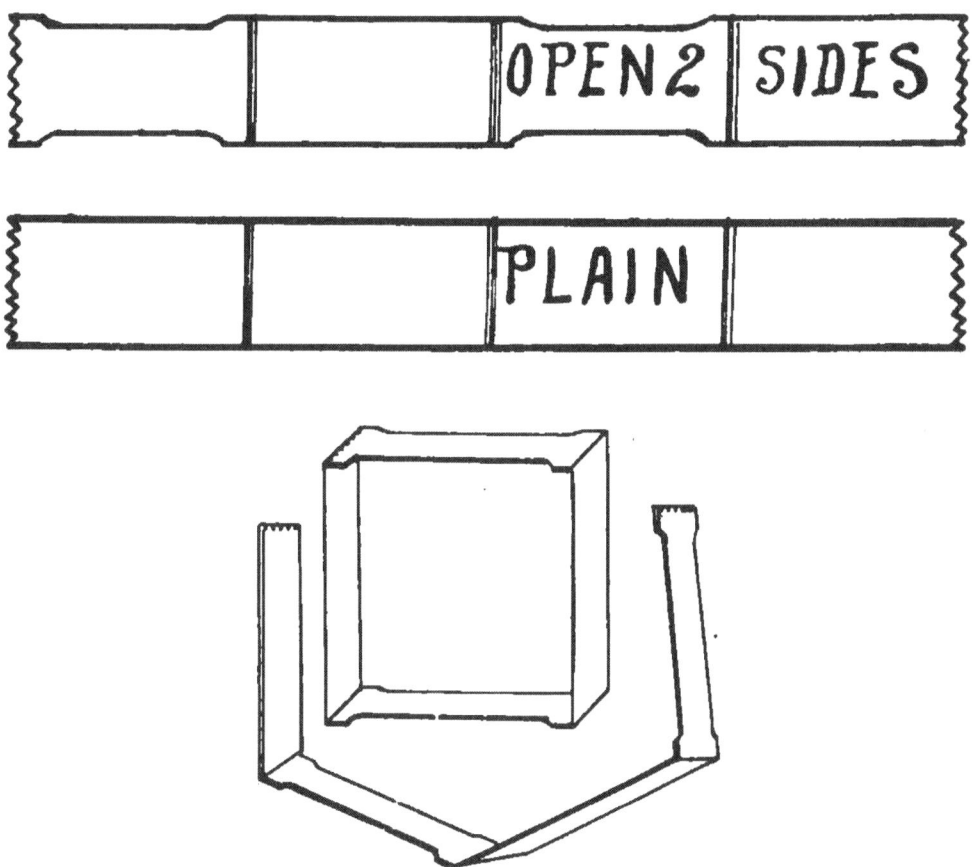

Fig. 70.—Sections for comb honey.

favorite in plain sections (Fig. 70). Honey in plain sections is a little more attractive when ready for the market than in the bee-way sections, as they can be scraped cleaner. The 4 × 5 plain section looks larger than the square section, although the weight may be the same.

Separators.—It is hardly within the scope of this book to describe the many different plans of utilizing sections for the purpose of obtaining comb honey in the small size quantities that the market demands. There are still several kinds of supers in use that are being displaced by more practical ones. Formerly a large size frame holding two rows, or eight sections in all, which could be placed in an ordinary hive body was quite gener-

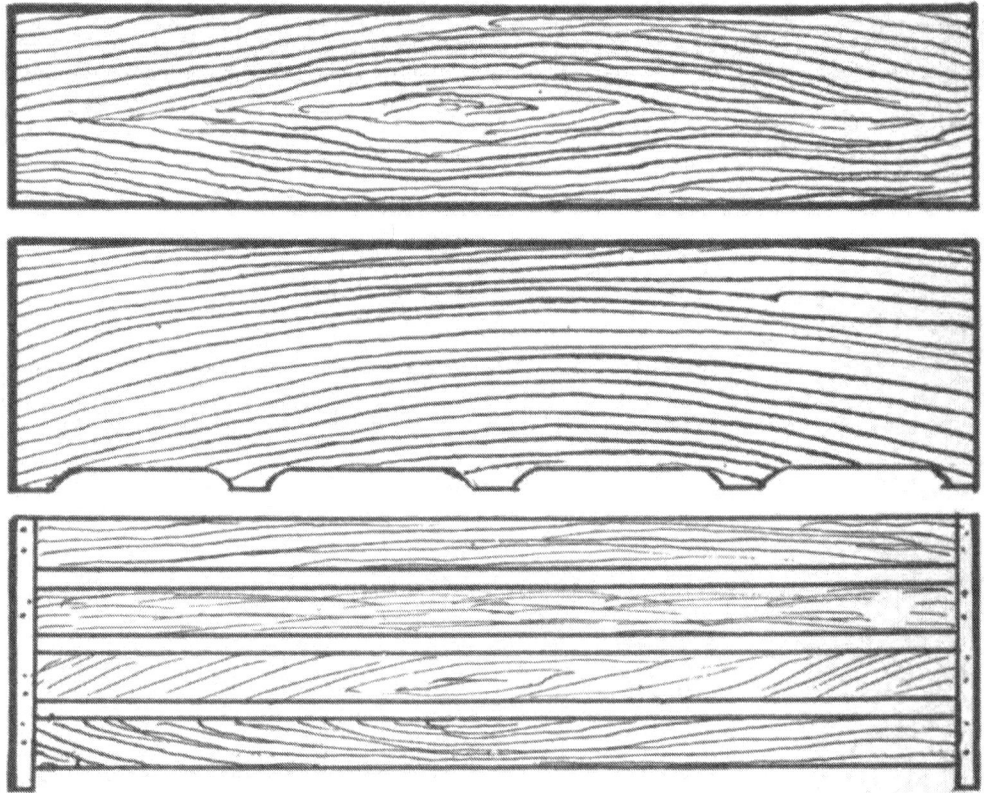

Fig. 71.—Separators for bee-way sections.

ally used. They have almost gone out of use and the super that holds a single tier of sections has been almost universally adopted. These can be handled to better advantage and the space added only as the bees are prepared to occupy it.

When the single tier super first came into use the sections were supported by strips of tin running across the under side. No provision was made for separating the sections, and as a result some would be very thick and some thin. Inasmuch as it is almost impossible to get sections filled of uniform weight this, too, has become almost obsolete.

It has been found necessary to make use of some kind of separator between the sections in order to get uniform results. Fig. 71 illustrates some different kinds of separators used. In the bee-way sections the passageway for the bees is cut directly in the section, while with the plain sections strips on the separators keep the sections a sufficient distance apart to permit the passage of the bees (Fig. 72).

Fig. 68 shows supers filled with the bee-way sections, while Fig. 69 shows a super for bee-way sections taken apart to show the separate pieces. At the top of the picture is a section holder

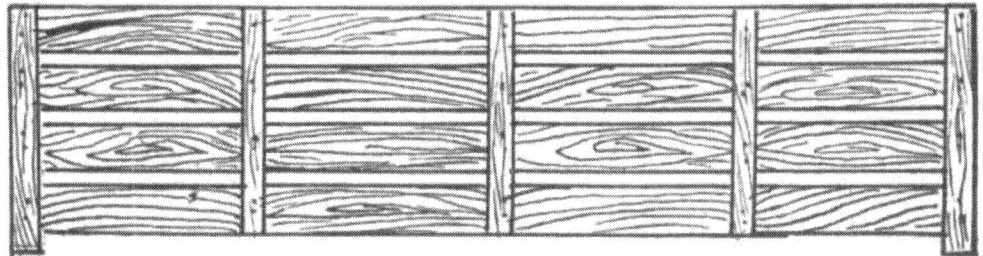

FIG. 72.—Fence for plain sections.

with four sections and on top of them an unfolded section lying against a folded one. Leaning against the super is a two bee-way section as they come from the factory in the flat. The other figures show the different types of separators for the two kinds of sections (Fig. 71). The separators commonly used for plain sections are composed of narrow strips that admit of the easy passage of the bees and are called fences. The bees can pass from section to section and from super to super much easier where the plain sections and fences are used than where the bee-way sections and solid separators are in use. The finished sections have smaller holes in the corners, which adds to the appearance of the finished article. The sections seem to be better filled also, as a rule in the plain sections separated by fences. When the slatted separators are used better results in the way of filling seem to be secured in the bee-way sections. Dr. Miller has discarded the plain sections in favor of the bee-way. This furnishes an argument hard to overcome since he is regarded as the most successful comb honey specialist.

Use of Split Sections.—For some reason the split sections do not prove popular although they have some advantages. The split section is cut through the center of three sides, thus permitting the use of full sheets of foundation which is attached by inserting a strip after a row of the sections is placed in the holder. The principal objection urged against them is the fact that the consumer, not being familiar with the methods of bee-keeping, may be suspicious of the wax that will show in this narrow crack around three sides of the finished section and conclude that the honey is manufactured. So many misstatements concerning honey production have been published that many persons are very suspicious and the split in the section with a showing of wax is not calculated to allay the suspicions of the skeptical person. Probably this fact has done more to prevent the adoption of the split section than all other reasons put together. If one sells largely in home markets where his product is well known and where he comes in close contact with his customers, this danger will not need to be feared, as in the case of honey sent to distant markets where no explanations can be made.

By the use of split sections much time can be saved in preparing the supers and putting in foundation. Four sections are filled at one time as shown in Fig. 73, which illustrates Dr. Leonard's method of preparing these sections for use. As will readily be seen by referring to the picture, four sections are placed in the holder, the opening slightly widened by the use of a metal form, and a sheet of foundation long enough to fill all four sections is slipped in. The foundation must be long enough to be caught at the ends to hold it firmly in place. The section holder is then removed from the form and the sections slipped in even with the outer edges of the holder and it is ready for the super.

By using split sections a nice appearing article is the result when finished, as there are no holes in the corners and if conditions are favorable the sections will be well filled.

Use of Foundation in Sections.—Comb foundation is pure beeswax rolled thin and by running between rollers printed with the size and shape of the bottoms of the cells. The use of foundation saves the bees much time at a season when every possible

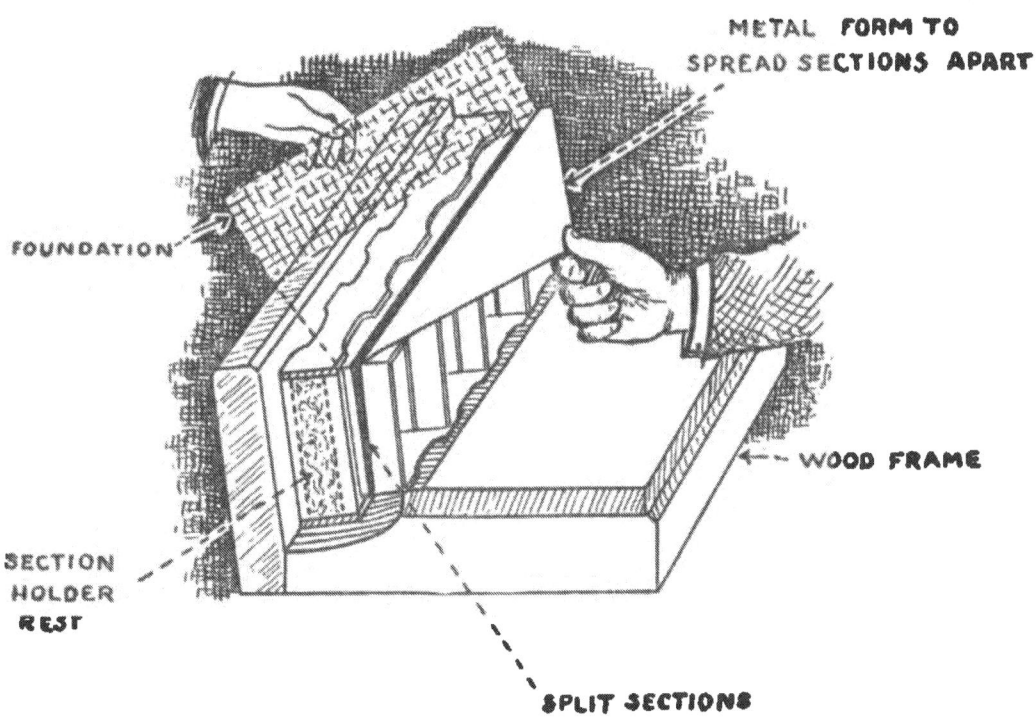

Fig. 73.—Dr. L. D. Leonard method of putting foundation into split sections.

assistance counts in additional honey stored. For use in sections only the thin or extra thin foundation should be used, as the thicker grades will be noticeable in the honey when taken into the mouth to eat. As only pure wax is used it is not an adulteration but if a thicker wax is used than the bees would build it will serve to make the product less desirable to the consumer.

While it is common practice among amateur bee-keepers to use but a small starter in the section, the extensive honey producer can ill afford to do with less than full sheets. When the honey flow comes it is important to make it possible for all bees to work and also important to save every unnecessary tax on their energy. In too many apiaries a single super with small starters will be placed on the hive and half the working force will be loafing for lack of storage room. Not more than a dozen bees are required to cover the bits of starter used by some.

The small compartments in which the bees are forced to

Fig. 74.—The Pangburn foundation fastener and sections filled with foundation.

work when storing in sections are unnatural, and considerable skill is sometimes necessary to get them started to work there at all. The small spaces make it impossible for the bees to cluster in large bunches as they do naturally when comb building. A dozen or two of bees will find it hard to reproduce a natural condition, but a full sheet offers much better opportunity. The wax which they need is already prepared to a large extent and a sufficient number can work together to assist in warming the wax and to encourage each other.

The method generally practised among large producers is to

use both top and bottom starters. The top starter is the full width of the section and lacks about half an inch of coming to the bottom. The bottom starter is about one-fourth inch in width. A wide starter at the bottom would fall over, while a small starter insures a well-finished section and that the comb will be attached at the bottom as well as at the top and sides. A small

FIG. 75.—Method of putting in foundation with Pangburn fastener.

space is left between the two, but this will be readily closed by the bees.

Putting in the Starters.—Various devices have been offered for fastening the foundation in the sections. If split sections are used, four sections will be filled as mentioned previously, and no special device for fastening will be needed as the sections will hold it firmly. With the ordinary sections some plan must be used for slightly melting the edge of the starter that comes in contact with the wood so that it will stick. Any supply dealer's catalogue will describe several of these devices. A metal plate about four inches wide and supported with a handle is probably as satisfactory and as rapid as any plan ever described. The metal is kept hot by an oil lamp and the edges of the foundation starters are touched with the hot metal as they are put in place in the sections. Fig. 74 shows a new plan for utilizing such a plate. This is known as the Pangburn fastener. (See also Fig. 75). A form is provided which makes it possible to place four sections in the holder and set them on the form. The four sections are filled at one time, thus making quite a saving in time.

If small starters are used a hot putty knife will serve very

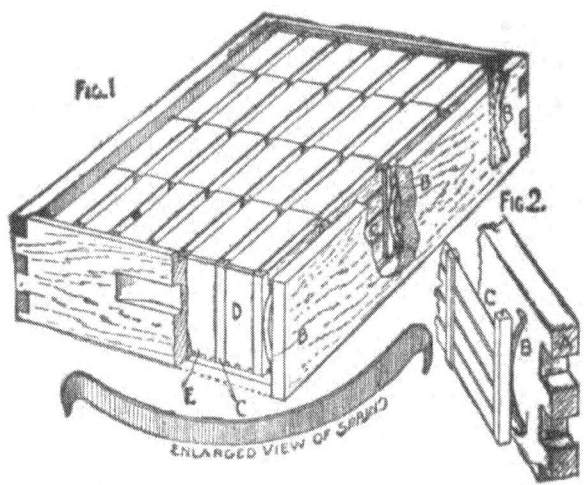

Fig. 76.—The use of super springs.

well. Although there are many devices offered by supply dealers nearly all operate on the same general principle. Which is best is to a large extent a matter of personal preference.

Super Springs.—When the sections are filled with starters and placed in the holders, and all are placed in the supers with separators between each row, there remains to fasten all together so tight as to make little daubing of sections necessary. For this purpose a follower board is used on one side of the super. Formerly this was fastened by means of a wedge which held all as tight as though made in one piece. However, when the sections are full of honey and the super is removed from the hive, it is not an easy matter to get them out without breakage. Super springs to replace the wedge have come into very common use (Fig. 76). One of these springs is placed at each end behind the board and answers all purposes nicely. The springs are very easily removed from the filled super and the follower board can then be pried loose and the operator has plenty of room to get the sections out. It is surprising to the novice how tightly the bees will seal every crack and crevice about the hive. In cool weather these fastenings hold as though they were glued, and provision needs to be made in advance to meet this condition. Super springs are regarded as a necessity by most comb honey producers.

THE SEASON'S MANAGEMENT

We come now to the most important part of the bee-keeper's business: the system of management. His hives may be of the best, all equipment may be the finest on the market, his bees may be of the best strain, and nectar may be present in abundance, yet if his system of management is not good his crop may be small.

At this point every bee-keeper must begin to be a law unto himself and to develop the system that best fits his locality and conditions. The most that an author can do is to make general suggestions as no system will suit all men and apply to all conditions. In a country like this of such vast distances, the flora will vary widely, the climatic differences are so great and other factors are so frequent that too many things need consideration to permit detailed directions. The best possible advice is to visit the nearest successful bee-keeper and learn as much as possible of his methods. Even a few miles often makes great difference in conditions that the bee-keeper must meet, so one must look for those things that are different in order to know how far the system will apply to his own conditions.

The bee-keeper needs to study general principles and to try to discover how they are affected by different conditions. Dr. Miller's great yield of nearly forty dollars per colony on an average was only secured after nearly half a century of study of the principles of bee-keeping as applied to a particular locality. True, conditions were exceptionally favorable and such an opportunity would only come once in years, but few bee-keepers would be so well prepared when it did come, or fully understand how to make the most of the favorable condition.

Prepare in Advance.—One great secret of success is in having everything in readiness when the flow begins—to have one's dish right side up when it rains honey. The winter months can be utilized to prepare a sufficient number of supers to care for any crop. The failure to provide supers in advance is common and one that costs the bee-keepers of the country thousands of tons of honey every good season. The bigger the harvest and the more urgent the need of extra room the less time there will be to prepare supers. Dr. Miller's advice is to have enough supers ahead to hold the biggest crop ever harvested in the locality and one extra super for each hive. After his big yield in 1913 he has not been heard to say how many that would take in his locality but prior to that time he estimated that at least seven supers for each hive should be ready to be safe (Fig. 67).

The fact that such a large number of supers will be left over from year to year leads most bee-keepers to neglect this precaution. If properly cared for they will not be injured even though not used for several years, and when the big yield does come they will be worth many times the cost.

Putting on the Supers.—Supers should not be put on every colony in a hit-and-miss manner whether they need room or not. Weak colonies that are not ready for storage room for surplus will be needlessly taxed to warm this extra space on cool nights and be further delayed in building up. Extra strong colonies will be ready for extra space before the average colonies, and the average colonies in turn will be ready some time in advance of the weak ones. There is no advantage in putting on supers when no honey is coming in, even though the colonies be strong. In most well-regulated apiaries one colony is kept on scales in order to ascertain when honey really is being stored and to tell something of the rapidity of the flow. This is a most excellent practice, as the apiarist can tell at once what really is being done.

A common question from beginners is how to tell when to place the supers on the hive. It is sometimes advised to put them on when the first white clover blossoms appear. This is perhaps as good advice as can be given if a definite time is to be set. However, so much depends upon the condition of the bees as well as weather and other conditions that no definite time can be set. The bee-keeper must come to know when his colonies have filled the brood chamber with brood and bees to the extent that they are ready to occupy the super and also to tell when they are getting something to put in it. No harm will be done in putting supers on strong colonies a few days before they are ready for them. It sometimes happens that some comb honey will be stored in supers by strong colonies from fruit bloom and dandelion, although this is not generally the case. If no supers had been supplied in a case like this there would be nothing left for the bees to do but to swarm.

Getting the Bees into the Super.—It is vitally important to get the bees to working in the supers as soon as possible to prevent the crowding of the brood chamber. They often hesitate to begin storing in the sections and sometimes will not do so at all without some extra inducement. One of the most common plans is to save all unfinished sections from the previous year to use as bait sections. Only the first super to be placed on the hive will need bait sections, as after the bees are at work in the supers they will occupy others as fast as needed.

If a sufficient number of these bait sections are at hand it is well to place one in each corner and one in the middle of the super. One in a corner at each end and one in the center will do very well. If the supply is short one in the center of the super will start them nicely.

For the purpose of starting the bees in sections it is also a common practice to use a shallow extracting frame at each side of the super. They will readily begin work in shallow extracting frames above the brood chamber, and once they have occupied these they will more readily occupy the section adjoining.

C. L. Pinney, who has secured splendid yields of comb honey, insists that by his method of combining the production of a small amount of extracted honey with comb honey he can get as many pounds of both as would be possible to get of extracted honey alone. Since his method has some features not generally practised, it may be of interest to describe it here. He has shallow extracting supers which he places on his hives when the comb honey is removed in the fall. These remain to catch any light fall flow and are left in place all winter, thus giving the bees a story and a half of comb surface on which to winter. In spring the bees will begin storing above the brood nest, of course, and as soon as the flow starts they will be at work in these shallow extracting supers. When he is ready to put on the comb honey supers he removes a partially filled frame and places it at each side of the comb super. There will then be two frames of comb with some honey in each comb honey super. The two frames are replaced with others that are empty and the extracting super raised up and the comb honey super placed underneath. The bees will, of course, go right on working on the unfinished extracting combs, and as soon as they are filled will fill the sections. When the comb honey super is well started it is raised up and another placed under it. The extracting super on the top may now be removed if desired, as it has served its purpose. The remaining honey may now be extracted and the frames put in a safe place until they are wanted to set on top of the hive when the comb honey is removed later in the season. If there is brood in the shallow frames it is his plan to use it for making increase. Several of the shallow frames of brood may be set in a super, and the super placed on the bottom board makes a shallow hive. By giving a queen, the little colony will build up nicely and when the frames are getting crowded may be placed over a full sized hive full of comb or frames of foundation.

When using this plan care must be taken to see that the queen is in the lower story when raising this shallow extracting super to place the comb honey super under it. It will usually be advisable to use a queen excluder (Fig. 77) under the supers to prevent the queen going above until the shallow combs are removed.

Essentials of Success.—The one big factor in getting a yield of honey, next to plenty of nectar in the fields, is to have big colonies of bees with the hive fairly running over with the honey

Fig. 77.—Ventilated bee escape and queen excluders.

gathering force. One strong colony at the beginning of the harvest is likely to store as much surplus as three or four of moderate strength and as much as a dozen that are weak when the harvest opens. From the time the honey is removed in the fall until the supers are placed on the hive for the next crop every move of the bee-keeper is made with a view of bringing the colonies to the next harvest with multitudes of bees.

Plenty of first quality stores and a large cluster of young bees insures good wintering, with proper care. The colony that comes through the winter with bees enough to cover four or more frames is the one that will build up quickly in the spring. Weak colonies are very slow in building up and the apiarist who does not fully master his business will have a large part of his colonies building up so late that the harvest will be half over before they are really ready for storing. Suitable spring protection, as discussed in the chapter on wintering, will have an important bearing on the condition of the colonies at the opening of the harvest.

Care of Weak Colonies.—Some bee-keepers take frames of brood from their best colonies in spring to give to the weak ones, thinking thereby to equalize the colonies and bring all to the opening of the flow in strong condition. Unless the stronger colony has seven or eight frames of brood this is not good practice. A better plan will be to take all very weak colonies and set them on top of the strong colonies, first removing the cover from the hive containing the strong colony and placing a queen excluder and a sheet of newspaper in its place. The queen excluder will keep the queens each in her own apartment and the paper will prevent the bees from fighting until they have become accustomed to the new condition and acquired a common odor or whatever it is that is characteristic of a colony and by means of which they recognize the numerous members of the same family. In a few days the bees will have made openings through the paper and the workers will mingle freely. Both queens will go on laying and the heat from the strong colony below will be of great help to the weak one above. In a few weeks they will have also become strong and may be again set back on their original stand. Two stronger colonies will result from a strong one and a weak one or even two weak ones, in this manner than by equalizing the brood and leaving them separate.

If a colony is sufficiently strong that the hive is getting crowded before time to put on the supers, it may then very readily spare a frame or two of brood to assist those which are not so far advanced.

Dr. Miller's Plan.—Since Dr. Miller perhaps holds the world's record of average production per colony, his system has attracted wide notice among the bee-keepers of the world at large. As before mentioned he uses the eight-frame Langstroth

DR. MILLER'S PLAN

hive. As soon as any colonies are strong enough to fill the eight frames he adds another story—a full sized hive body full of empty brood combs. Instead of putting this empty story on top he raises the hive and places it underneath. In this manner the heat of the colony is fully conserved. At the same time the bees will work down as fast as they need the room. White clover is the chief source of his surplus, and he endeavors to keep the bees occupied with breeding until the beginning of the clover flow.

When conditions are right for putting on the supers he again reduces the colony to eight frames. If there is less than eight frames of brood he places it all in a single hive body and places the super on top to provide the room formerly given by the extra hive body. If there are more than eight frames of brood the extra frames are given to colonies with less than eight frames. If, as sometimes happens, he has some frames of brood left after all colonies are provided with eight frames in single hive bodies, the rest is used to make increase, or to form nuclei or is even placed in hive bodies which are piled one on top of another to permit the brood to hatch, and latter be used where needed. If there is no other use for it a queen is given or else one is raised from young larva in the hive and a strong colony is the result.

Concerning additional super room Dr. Miller says:

During the early part of the harvest, so long as there is a reasonable expectation that each additional super will be needed, the empty super is put under the others, next to the brood chamber. Work will commence in it more promptly than when an empty super is placed on top, and that greater promptness in occupying the new super may be the straw to turn the scale on the side of keeping down the desire for swarming. But when a super is put on toward the close of the season, not because it seems really needed but as a sort of safety-valve in case it might be needed, I do not wish to do anything to coax the bees into it, so it is put on top, and the bees can do as they please about entering it. It is true that if an empty super is placed under the others at a time when the harvest is nearing its close, the bees may not do a thing in it, but merely go up and down through it and keep to work in the super above. But it is not so well to have them working so far from the brood nest with empty space beneath.

Latterly I have fallen into the habit of giving an empty super on top, even when an empty super is put under. The empty super on top gives a less crowded feeling and may help a little toward preventing swarming. No matter how full or empty the lower super may be, this top super serves as a sort of safety-valve, in case any need for more room should arise. The next time there is need to give a super below, this top super is moved down and another empty super put in its place. When the top super is put down, I think the bees start work on it just a bit sooner than if it had not been above.

The ability to provide the bees with sufficient room to make the most of the harvest and yet keep them sufficiently crowded to get the sections well finished is the great test of the scientific bee-keeper. If, when the honey is coming in with a rush, too many supers are put on, the bees will scatter their forces to such an extent that when the flow stops there will be a lot of unfinished sections which cannot be marketed. On the other hand, if insufficient room is provided there will be a loss of honey for lack of storage room.

Nothing can take the place of experience in determining this matter. With the beginner it will be almost altogether guess-work and as Dr. Miller says the guess-work will never be altogether eliminated, for no man can tell ahead how long a honey flow will last.

SWARM CONTROL

The man who will find the secret of swarm control will confer a great blessing on the fraternity and his name will not be forgotten. Many and various are the plans recommended to prevent swarming. While the extracted honey producer is able to reduce this trouble to the minimum and in ordinary seasons have little difficulty, the comb honey producer who does not find it his greatest problem has not been heard from. A strong colony may swarm and take enough bees to found a new colony and store a profitable crop, while at the same time leaving enough force on the old stand to store some surplus. It will sometimes happen that about as much honey will be stored by the two divisions as would have been the case had the colony not swarmed.

In localities where the flow is sufficiently rapid to make comb honey production profitable, it is likely to be short and the colonies which do not swarm are usually the ones to store the big crops. At any rate the bee-keeper prefers to make increase at his convenience and not to be watching for swarms all summer. In a large apiary where there is no control of swarming there is little time for anything else than watching for swarms and getting them hived.

Breeding to Produce a Non-Swarming Strain.—In spite of the fact that several writers, notably Dr. Bonney, take the position that the honey-bee cannot be improved because of the difficulty of controlling male parentage, much is to be hoped for along this line. Even now some progress is being made and a few leaders among the enthusiasts who are persistently following up the method of selecting the best honey producers among the non-swarming colonies and rearing queens only from them are getting results. It is true that progress is slow and that discouragements sometimes are to be met, but some claim a noticeable decrease in the number of swarms as a result of such breeding for a series of years.

Experiments looking toward the artificial mating of queens have been made from time to time with uncertain results. Once let a satisfactory method of accomplishing this be found and the great problem of breeding good bees is solved. As long as the queen must mate in the air according to the natural provision she may mate with any one of a thousand drones that chance to be flying at the time she takes her marriage flight. If a method of safe artificial fertilization can be devised this uncertainty is removed and drones from the best colonies can be selected. It will then be an easy matter to breed from stock showing any particularly desired trait and as good results can be expected as have resulted from similar efforts to improve live stock and poultry. The non-sitting breeds of fowls are pretty good evidence that it is possible to breed out even the strong natural instincts. In a state of nature the sitting of the hen was essential to the perpetuation of the race. The invention of the incubator removed the necessity for sitting and the poultrymen proceeded to remove the tendency to sit on the part of the hen, by natural selection. In a state of nature only a few eggs were laid but now behold the two hundred-egg hen, the result of the effort of the breeder.

Control of the male parentage is the only obstacle in the way of the bee-breeder and in spite of that he is accomplishing something. Not long since the result of an experiment along this line by Professor Francis Jager and an assistant in the University of Minnesota, which was apparently successful was published in "Science." The queen emerged from the cell with only rudimentary wings and was never able to fly, hence there could seem to be no mistake in this case. The bee-keeping world is watching with interest the progress of further experiment along this line in the hope that fertilization by artificial methods may some day be possible.

Cutting Queen Cells to Prevent Swarming.—Perhaps the cutting of the queen cells was the first method devised looking toward swarm control. It is probably the method most commonly practised. Yet it is not entirely dependable. In some instances if the cells are removed as soon as the larvæ first appear in them, no more will be built and there will be no swarm for that year. If, however, cells are once sealed and the bees have the swarming fever, they will build one batch after another until the bee-keeper will find it cheaper to let them swarm and be done with it, than to examine every comb and remove royal cells every ten days all summer. Occasionally one will be missed and then out comes your swarm whether or no.

De-queening During the Honey Flow.—A few bee-keepers go through all their colonies during the honey flow when it is expected that eggs laid will not mature in time to be of assistance during the harvest and kill all the queens. At this time queen cells will be built in many colonies in preparation for swarming. All cells will be cut out at the same time. Weak colonies or others not likely to swarm are passed, as are also any favorites from which it is desired to get cells for making increase or similar purposes. The colony being queenless will at once build several cells in order to provide another. Some system of marking is used to note the condition of the colony. Nine or ten days later a second trip is made through the yard, to cut out all queen cells but one in each hive. At this second visit all cells found in the colonies marked as A. No. 1, will be saved. Only one will be left to insure a queen and the others will be placed in the hives which have markings showing that they are not up to the standard. All poorer or surplus cells are destroyed.

An accident to a cell or to the new queen on her mating trip would leave the colony hopelessly queenless, as there will no longer be eggs or young larvæ in the hives. To provide against such contingencies a number of nuclei are started and provided with cells to insure a sufficient number of extra queens to supply the colonies whose queens are not successfully mated. A third examination will be necessary after the elapse of a similar period to ascertain whether queens are present and to supply those colonies where failure has resulted.

It will sometimes be necessary to cut cells from a part of the colonies in advance of the time of this wholesale de-queening, or as only a small number of swarms will issue they may be hived in the usual manner. This method, while somewhat drastic, has the desired effect and perhaps comes as near controlling swarming as any other. In the discussion that followed the presentation of the plan by F. W. Hall at the Iowa Bee-keeper's Convention there was serious objection to it on the part of some very successful apiarists. It is contended with good reason that many valuable queens will thus be destroyed and that one year is not long enough to give a queen an opportunity to show her good points. Those who follow the method, it will be noticed, make exceptions of those queens which are especially promising and retain them as breeders. While there are some extensive bee-keepers who will find the method suited to their requirements, it is not one to be generally recommended under all conditions.

Dr. Miller practises a de-queening treatment along different lines. He removes the queen and places her in a cage where she is cared for by her own bees, or she is introduced in a nucleus where she continues to be busy. Of course all cells are destroyed or removed at the time the queen is taken away. At the end of ten days the cells are again removed and the old queen returned to them or another given in her stead. This is the same treatment in effect excepting that he retains his queens as long as they give satisfactory results, whether for one year or three or more.

With the exercise of the utmost care there will be plenty of swarms some seasons while other years the matter can be controlled without great difficulty. The swarming tendency can, however, be so far checked as to greatly increase the returns from the apiary.

Space Under the Brood Nest.—An empty space under the brood frames seems to serve to some extent the purpose of swarm prevention. A deep bottom is better than a shallow one. In Europe the Simmins plan of placing the comb honey supers with empty sections under the hive, to begin with, seems to be practised to some extent, although the author does not know of its use in this country. According to this plan empty supers are kept in place under the hive all through the honey flow. The bees prefer to store their honey above the brood nest and will do so if possible. When they are getting too crowded above they will begin to build combs in these comb honey supers below in preparation to working down. It is now time to remove them and place them on top of the hive and put another empty one in place underneath. This empty space below serves as additional clustering space and also facilitates ventilation.

A somewhat similar purpose is served by the usual practice of blocking the hive up at the corners during the honey flow if the weather be hot. The hive being open on all sides there is free ventilation, which is a material factor in swarm control.

BEE ESCAPES

REMOVING THE HONEY FROM THE HIVE

As soon as possible after most sections in a super are nicely capped the honey should be removed from the hive to prevent travel stain. If the bees are forced to pass over the sections in going to and from the supers above for any considerable length of time, the white cappings will become discolored and the market value be reduced. True, the bee-keeper tries as far as possible to have the sections finished in the top super so that there will remain no necessity for much travel over it when capped. It is not always possible to arrange the supers in the ideal manner and even if capped in the upper super some travel stain will result if the honey is allowed to remain too long. Comb honey in sections is usually sufficiently ripened by the time all but the corner sections are capped.

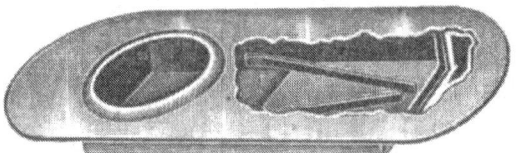

Fig. 78.—The Porter bee escape.

Bee Escapes.—The invention of the bee escape was a great boon to the comb honey producer. The Porter escape is the one in most common use (Fig. 78). A board the size of the hive has the escape fitted to a small opening in the center. The construction is such that the bees can go down through it but cannot return. By putting on the escapes in the evening it is usually possible to remove the supers of honey, free from bees, the following morning.

Some bee-keepers depend upon driving the bees from the supers with smoke, but this often results in injury to the honey, as the bees will uncap the cells in order to reach the honey. Unless some care is used the combs may be somewhat discolored also. The escapes are inexpensive and the small outlay is more than repaid in the saving of time alone.

The LaReese or ventilated escape has some advantages over

the other (Fig. 77). This is made with about one-third of its surface covered with a double screen. Double wire cones make it possible for the bees to go down easily but difficult to find their way back. The principal advantage of this excluder over the other is the ventilation, which prevents the melting down of the combs in extremely hot weather. While this happens rather infrequently, conditions occasionally are such that honey will melt badly in unventilated supers. Over these ventilated escapes the air will be kept moving by fanning bees below, even if none are left in the super, and the bad effects of the heat will be avoided.

If it is desirable to remove the super before the outside rows of sections are finished they may be set aside and replaced in other supers.

Closing the Season.—Unless there are unfinished sections which may be left on the hive in the hope of finishing during the fall flow, it is usually advisable to remove the comb honey supers at the close of the main harvest and to replace with extracting combs to catch any light fall flow. In localities where the fall flow is such that comb honey can be produced profitably this will not apply. In many localities the fall flow is so light that the sections will not be well finished and if marketable at all few of them will grade better than number two. Unless one can produce comb honey of the best quality it is better to have it stored in the extracting combs.

When one comes to remove honey in wholesale quantities after the close of the flow it will be necessary to proceed carefully or there will be much annoyance from robber bees dodging into the sections and flying home with a load. Let a few bees get away successfully in this manner and shortly the air will be full of bees intent on finding the source of supply. At such times it becomes very difficult for the operator to work and there is greatly increased danger from stings. Everything should be kept closed as carefully as possible and when the supers are removed they should be covered at once.

For convenience at such times most bee-keepers have strips of canvas or muslin, large enough to cover a pile of supers, which are called robber cloths. It is well to remove the supers at once to the honey house where they will be safe from visiting bees. The honey house should, of course, be so tight that no bee can find its way in, but with escapes at the tops of the windows to make it easy for any chance bees to get out. In bringing in honey from the apiary it will frequently happen that a good many bees will still remain in the supers. If the house is properly constructed they will make but little trouble as they will fly to the windows and escape. (See Honey House in next chapter.)

Removing Sections from Supers.—The super springs already described make it easy to loosen the follower board which will give room to work. Each section holder may now be crowded over into the vacant space and removed with its sections. A better way is to push the whole lot out at one time. There are two ways of doing this. Either have a form the size of the inside of the super and set the super on it; with a mallet or other object drive the super down outside of it, leaving the section holders and their contents on the form; or have an empty super on which to set the filled one upside down. Then by carefully jarring the section holders they may be pushed down into the empty super. As soon as loosened fully from the propolis and wax they may be lifted out. A little experience will greatly facilitate matters in thus removing the sections. As a rule the novice will break a number of sections before he learns how to proceed without injuring the honey.

After the sections are taken from the supers they should be sorted and all unfinished sections replaced in the supers to be replaced on the hives to be finished if the season is not too far advanced, or set aside to serve as bait sections next season.

Fumigation.—Unless the season is so far advanced that freezing weather is at hand, some precaution will be necessary to insure that none of the crop is spoiled by wax moths in storage. At any rate comb honey should not be long subjected to freezing temperatures because of the danger of granulating. If the honey is to be sold at once no harm will be done by fumigating, as one would not wish the buyer to wake up to the fact that wax moths were destroying his honey a few weeks later.

Eggs may be present even though there is no appearance of moths when the honey is taken from the hive. An occasional examination will reveal their presence when they may be destroyed by the usual methods of fumigation.

QUESTIONS.

1. Under what conditions is the production of comb honey satisfactory?
2. Discuss hives for comb honey production.
3. Describe the different sections used for comb honey.
4. Note the advantages of split sections.
5. Discuss the use of foundation in sections.
6. What advance preparation should be made for the season's work?
7. When is a colony ready for supers?
8. What methods are used to get the bees into the supers?
9. What are some of the essentials to success in comb honey production?
10. How may weak colonies best be built up?
11. Outline Miller's method of comb honey production.
12. Discuss swarm control.
13. When should comb honey be removed from the hive?
14. Of what advantage is a bee escape?

CHAPTER X
PRODUCTION OF EXTRACTED HONEY

WITH proper equipment, extracted honey production is a pleasant and profitable pursuit. Without it, it is dirty, mussy and disagreeable. Less skill and labor may perhaps be required in specializing in extracted honey. If the market is properly developed, it may be as profitable or more so than comb honey. As generally handled, much more extracted honey will be produced than comb honey, but skillful apiarists who know how to make the most of the opportunity will get very nearly as many pounds of comb honey as extracted where honey flows are very rapid. If one wishes to do business on a large scale, and to run a series of out apiaries, there are less difficulties to be overcome in the production of extracted honey.

Proper Equipment.—The kind of equipment that will be needed will depend much on the extent to which one wishes to develop the business, and whether one plans a central extracting house, where all honey is brought to be cared for, or whether one uses a portable outfit with a small honey house at each apiary. Which is the better plan, the author is not prepared to say, for there are extensive honey producers some of which prefer one and some the other.

In any case the extractor is an important article. Larger extractors can be used in the central plant than are practicable to carry from place to place. For portable outfits, the four-frame reversible extractor is usually used. For a small home apiary, a two-frame extractor will do very well, but if there is any idea of extending the business, nothing short of the four-frame capacity should be bought.

Extractors.—Until the invention of the extractor in 1865, the nearest approach to extracted honey was strained honey. This was a common method until but a few years past. Surplus honey was removed from the hive by cutting out the combs and

Fig. 79.—Sphuler's hand extractor as used in Europe.

EXTRACTORS

mashing them up in a cotton cloth which was hung up in a warm place to drain. Masses of brood, pollen, and honey were often broken up together, so that the quality was anything but attractive. Many people who have not kept pace with the progress of bee culture, seeing extracted honey in the market, refer to it as strained honey.

The frames full of sealed honey are now taken from the hive,

Fig. 80.—Storage tanks of a large honey producer in California.

and by means of a warm knife the cappings are skilfully cut from the comb. The frames are then placed in the basket of the extractor (Fig. 79), and the machine started. The centrifugal motion throws the honey from the side of the comb next to the side of the can. The machine is then reversed, throwing the honey from the other side in the same manner. The honey is drawn off in tanks, or in smaller containers, according to circumstances (Fig. 80).

Since the first extractors appeared in the market, many improvements have been made. The first extractor revolved altogether, tank and all. Then came an extractor in which two comb baskets revolved inside the can, but requiring that the combs be lifted out and turned around, after one side had been emptied. The latest machines are reversed by the simple pulling of a lever without stopping the machine. The larger sized ones have a

Fig. 81.—A power driven extractor.

capacity of eight frames, so that something like forty pounds of honey is extracted from a set of full combs at each operation.

Power.—For extensive apiaries, the power driven extractor (Fig. 81) is a great economy, for while the operator is uncapping one set of combs, the machine will empty another. A small gasoline engine costing from thirty-five to fifty dollars is sufficient to furnish the necessary power, and, during the extracting season, will nearly take the place of one man, and at much less cost. The labor item is the heaviest expense with most lines of productive enterprise, and any machine that will reduce this expense will add materially to the net profit at the end of the season.

The same power can be made to serve for many other purposes, such as pumping water, running the washing machine, cream separator or other small machinery. The gasoline engine is generally regarded as a necessity in the apiary, unless it be within reach of electric power.

Honey Pump.—The honey pump is a comparatively new invention and has not, as yet, come into general use. Whether its use will be advisable will depend a good deal upon the construction of the honey house. (See Honey House.) If the storage tanks are on a level with the extractor or above it, the honey pump will be a time saver in the large plant. In the past these new machines, like most new inventions, gave more or less trouble in their operation. The machines are now perfected to the point where they are run with good results. The pump is attached directly to the extractor, and run by a belt attached to the reel of that machine. The same power runs both and the honey is pumped into the storage tank as fast as extracted. This not only saves the labor of handling the pails of honey as drawn from the extractor but relieves the care of watching for fear the pail will be neglected a moment too long, and the honey run over and be wasted. The extractor can also be fastened directly to the floor, instead of upon a platform, as is necessary where provision must be made for a container under the honey gate of the machine.

Storage Tanks.—Tanks of sufficient capacity to hold the season's crop should be provided, for it is not always advisable, even if there is time, to get a part of the honey to market during the season. Many bee-keepers provide a sufficient tank capacity to hold the output of three or four days' extracting, and have on hand a large number of sixty-pound cans in which to store the bulk of the honey. It is drawn into the cans as soon as it has settled a few hours, and is then ready to ship to a wholesale market, or with the cans tightly closed is safe from dust or spilling, if the honey is later to be sold in small containers (Fig. 82).

Galvanized iron tanks are quite commonly used for storage, though some use wood tanks for this purpose. In some localities barrels are used, but this is no longer common in many localities in the United States. Honey should not be left long in open

Fig. 82.—Sixty pound cans for extracted honey.

tanks for reasons of cleanliness. It is thought also that honey exposed long to the air loses some of the delicate aroma peculiar to the finer grades.

Uncapping Boxes.—Several different kinds of uncapping cans or boxes are in the market, and many more are in use in the apiaries. The accumulation of cappings during a week's extracting will be surprising. It is not only necessary to save these cappings for the wax they contain, but much honey will be carried with them which, also, is to be cared for.

A good uncapping can provides for the draining of the cappings, so that the honey will separate from them as rapidly as possible. Some of the boxes made for this purpose have slatted

CAPPING MELTERS

bottoms, through which the honey is drained into a tub or pail set for the purpose (Fig. 83). The cans are provided with coarse screen, which catches the cappings but permits the honey to run through in the same manner. Some of the best of these are made at the apiary where they are to be used, thus fitting the available space in the honey house, and meeting the individual requirements of the bee-keeper. The uncapping box or can should be of a convenient height and have a suitable rest for the comb when the cappings are being removed.

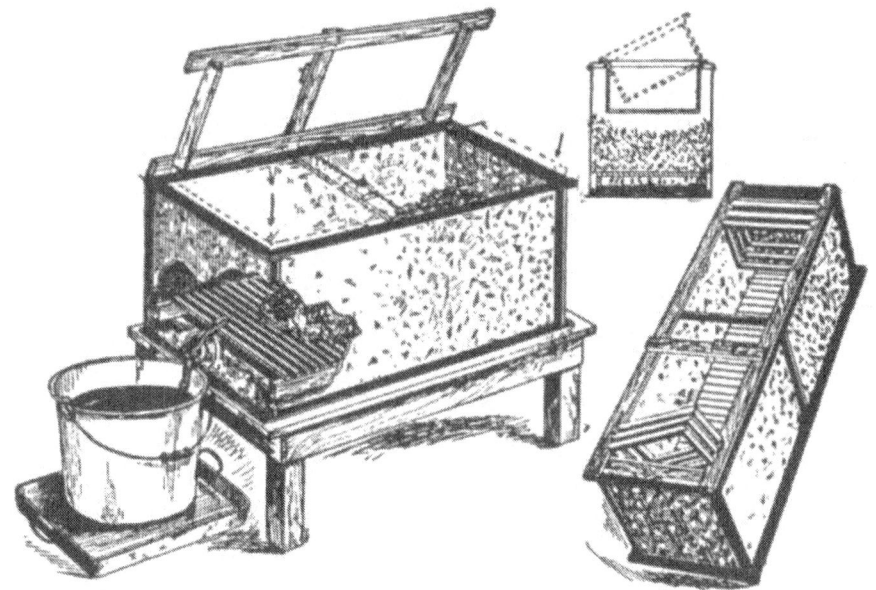

Fig. 83.—The Townsend uncapping box.

If the box is made rather long, and the width is the same as the length of the extracting frames, they may be left hanging in the box as fast as uncapped until removed to the extractor. In this way the box will catch the drip from the uncapped combs. Something similar to the Townsend uncapping box shown in the illustration (Fig. 83) is probably most commonly used.

Capping Melters.—The capping melter is somewhat similar to the uncapping box, but has a sloping metal bottom. Under this is placed a small oil stove, or some other device for heating the metal bottom (Fig. 84). The idea is that as fast as the cappings drop on the hot metal, they will be melted and run at once into a receptacle at the lower end. The bottom is usually double, and the metal extends up several inches on each side of the tank, the space between the two sheets of metal composing the bottom and sides being filled with water. The water distributes the heat more evenly, thus preventing the space immediately over the flame from becoming too hot.

The honey and melted wax are both caught in the same container, but the wax comes to the top and may be lifted off in cakes when cool.

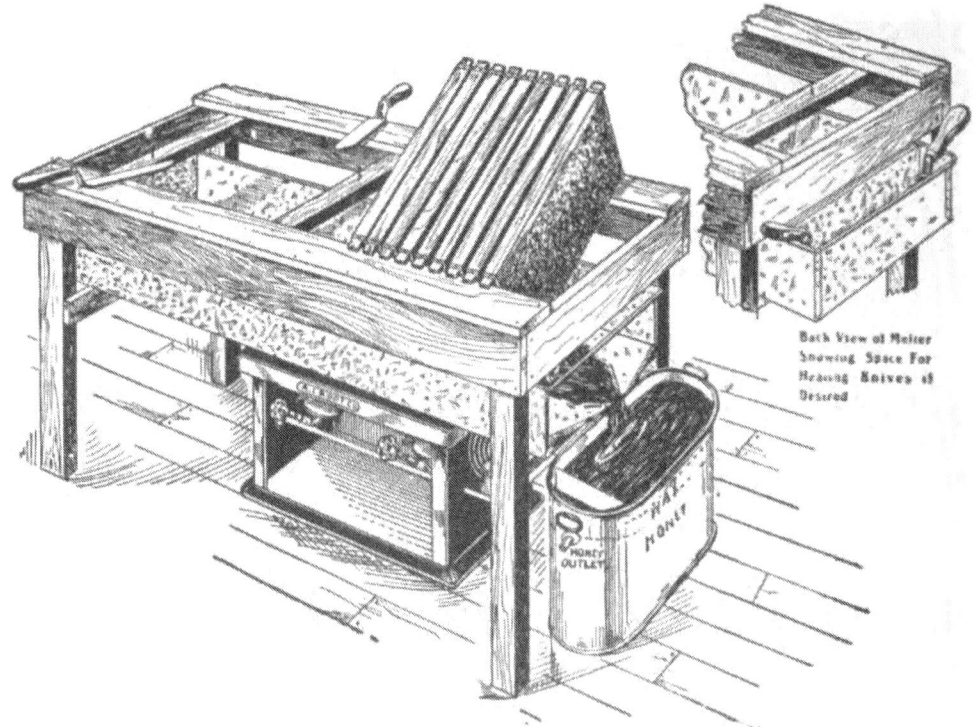

Fig. 84.—The Peterson capping melter.

There is quite an advantage in disposing of the cappings as fast as cut from the combs, especially in a large apiary. However, these melters do not always give satisfaction, as the honey is often over-heated and the quality injured. Most large producers of the author's acquaintance prefer the uncapping box without artificial heat.

Uncapping Knives.—To remove the cappings a knife with long blade is used. Straight knives were formerly in general use, but of late the Bingham knife has largely replaced all others. Fig. 85 shows this knife. For use it is kept hot by dipping in hot water.

A somewhat similar knife heated by steam is used to some extent in large apiaries. The steam knife is hollow, and is attached by rubber tubing to a small vessel of water which is set over the burner of a small oil stove or lamp. A small hole at the end of the knife permits the escape of the steam. As the temperature of the knife is evener, it is more satisfactory. The cappings do not stick to the knife, or the operator does not have to stop frequently because it has become cold.

Fig. 85.—Bingham uncapping knife.

Hives for Extracted Honey.—While there is a difference of opinion regarding the size of hive most profitable for the production of comb honey, the bee-keepers are nearly all agreed that the large hive is the thing for extracted honey. The ten-frame Langstroth is more generally used for this purpose than any other (Fig. 86). Some extensive producers use a twelve-frame hive with the same size frames (Fig. 87).

The Dadant hive is very satisfactory for this purpose, and were it not for the fact that the Langstroth frame is in more general use, it would find favor with the producer. The brood frames of the Dadant hive are too large, however, for extracting frames, thus requiring two sizes of frames. It is a decided advantage to have all frames in the apiary of the same size and style.

Shallow extracting frames are good for use in connection with the comb honey apiary, but are not to be recommended for the production of extracted honey. Nearly as much time will be required to uncap a small frame as a large one, and extra time will be required in the manipulations, which is decidedly to their disadvantage.

Not only should everything about the apiary be planned to avoid the loss of time, but the accumulation of unnecessary equipment should also be avoided. If extracting frames and brood frames are of the same size, extra frames may be used for either purpose for which they are needed, instead of having to supply more when one or the other runs low. It might easily happen

Fig. 86.—Langstroth hive for extracted honey.
Fig. 87.—Langstroth hive dissected.

when there are two sizes that there will be a surplus of one at hand and a lack of the other.

Kind of Frame to Adopt.—As to the kind of frame which best serves the purpose, there is a decided disagreement. While the author personally prefers the Hoffman self-spacing frame, most of the large producers seem to be decidedly of the opinion that a loose-hanging frame is better.

The Hoffman frames (Fig. 93) require little attention to see that they are properly spaced when placed in the hive. On the other hand the loose-hanging frames must always be spaced after setting in place, or the combs will be unequal in thickness.

There are several devices for spacing the frames. Staples are used in some cases. The metal spacers are very popular also.

These are so placed near the top of the frames that they will be the proper distance apart when placed in the hive.

While there may be a difference of opinion regarding the best, from the standpoint of the extensive extracted honey producer, the novice will find the self-spacing frames much better, as there is less danger of getting the brood nest too crowded or the combs too far apart.

Some contend that the use of metal spaced frames tends to dull the uncapping knife by constantly knocking against it. This argument carries little weight, for a good operator will seldom strike the edge of his knife against the metal.

THE HONEY HOUSE

A good honey house is a necessity in extensive honey production. The small honey producer can get along with a large room in the dwelling house if necessary, but the nature of the work of extracting is such that a separate building is very desirable. It need not be expensive, but must be tight enough so that no bee can enter when doors and windows are closed. If the bees once find their way in when a lot of honey is exposed, they soon come by thousands and make work impossible. During a good honey flow they are so busy bringing in nectar from the field, that they pay little attention to anything else. At such times extracting can often be done out of doors without annoyance. A check in the honey flow brings a decided change in their attitude, and they will soon be seeking every possible opening to a building where honey is stored.

If portable outfits are used and the honey extracted at the various apiaries, small buildings will serve very well, because the honey will be taken away as fast as extracted. It is a common practice among bee-keepers following this plan to visit a yard in the morning and spend the day extracting, and take the honey home at night.

Even though the portable outfits are used, a good-sized building will be needed at the home yard where the honey is prepared for market and the various appliances prepared for use.

The illustrations show two good kinds of honey houses. Fig. 88 shows a honey house two stories high. This house has some decided advantages, and, although it was built at a cost of about

Fig. 88.—A well-arranged, two-story honey house.

one thousand dollars, the extensive honey producer will find it well worth the cost. By looking at the picture it will be seen that the lay of the land is such that the ground is on a level with the floor of the upper story at one side. At the other side the ground is on a level with the lower story. A side hill location is not always available, and otherwise this two-story arrangement would not be very satisfactory, as too much energy would be necessarily expended in getting the honey upstairs. In a case like this, however, the honey can be unloaded on the upper floor without extra effort.

On the upper floor is the power driven extractor. From it there is a pipe leading directly to a large settling tank on the floor below. The honey will thus never be handled from the time the uncapped frames are placed in the extractor until it is drawn

Fig. 89.—Large honey house with all work on ground floor.

into the sixty-pound cans to ship to market. This particular honey house is arranged with the idea of eliminating every possible unnecessary item of labor. One man has produced, extracted, and prepared for market something like forty thousand pounds of honey from five yards, with help only a few days during the busiest season.

On the upper floor is the work shop, where hives and frames are assembled, and where extracting combs are stored, in addition to the extracting room. On the lower floor is the big settling tank, the bottling room and storage room for honey. A better arranged or more satisfactory honey house could scarcely be planned. The honey room must always be kept dry to avoid injury to the honey.

Fig. 89 shows another kind of honey house. Here everything is on the lower floor, excepting storage for unused equipment. The building is composed of three large rooms. At one end is the general storage room. In the center is the extracting

Fig. 90.—The automobile is valuable for outyard work.

room, also used for preparing equipment, wiring frames, etc. At the other end the automobile is driven in with the load. The automobile is a very useful, and now almost necessary adjunct to a large apiary, where outyards are widely scattered, as the time saved in travelling to and from the yards is an important consideration (Fig. 90).

As will be seen from the two pictures, the extensive production of honey necessitates a large building for comfortable work. The tendency is always to build too small, and crowding does not tend to economy of time or labor. If the beginner who expects his business to grow will plan his honey house so that additions are easily built on, he will be wise.

Floor.—A cement floor is very desirable, as it is much easier to make the building proof against rats and mice. Neither should be tolerated in the honey house, as they are the source of great annoyance and damage. Mice will destroy many dollars worth of extracting combs, unless they are stored beyond the reach of the rodents. A cement floor also makes a better foundation for fast-running machinery.

Doors and Windows.—The windows should be tightly screened to keep out flies and bees, but the doors are better without screens. If the doors are screened, they will be left open when the extracting is going on, and large numbers of bees are likely to collect on the screens in an effort to get in. Every time the screen is opened a few of them will dodge in, with the result that a constantly increasing number are flying about, which is annoying to the bee-keeper and bad for the bees. If only the windows are screened, the doors will be kept closed excepting when necessary to pass in or out, and the bees collecting on the outside will gather at the windows where they will be unable to enter.

Escapes.—Bees that are carried into the honey house will naturally fly to the windows in an effort to escape. At the top of every window should be provided an escape which will permit them to get out, but which will turn those on the outside which may try to get in.

A good method is to place strips of lath under the wire screen, thus holding them out a quarter of an inch from the building. If these strips extend about six or eight inches above the top of the window, and the screen extends as far, the space under the screen may be left open at the top. The bees on the inside will walk up and out, while those outside will not go much above the window opening and will not find their way in.

Another method is to leave several wrinkles in the screen

along the top. Each of these places will leave an opening large enough to permit the bees to find their way out. To prevent those from the outside from coming in, wire cones are placed over the openings.

Still another common plan is to place ordinary bee escapes,

Fig. 91.—Comb at right built on full sheet of foundation; at left, without foundation.

such as are to be purchased from any dealer in supplies, in the corners of each window. This plan does not work well in practice. Any method that will permit bees to go out without letting outsiders in, will be satisfactory.

PREPARING FOR THE HARVEST

The importance of having combs built on full sheets of foundation to prevent the building of drone comb is mentioned elsewhere. Drone combs are not especially objectionable in extracting supers, as long as the queen does not have access to them. The productive bee-keeper, however, should avoid having them built in the first place, as they should never be permitted in the brood chamber, and, unless excluders are used, the queen will sometimes be laying in the extracting supers. It is highly desirable that every comb be so perfect that it can be used in any part of the hive for any purpose needed.

Aside from the necessity of avoiding the drone comb, it is

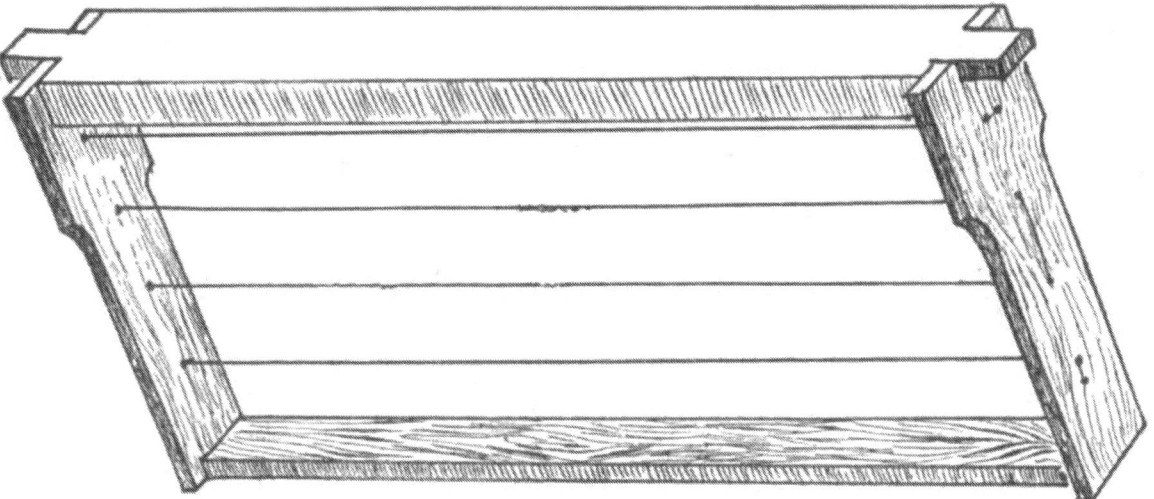

FIG. 92.—Usual method of wiring frames.

possible to get so much better combs by the use of foundation, and to have them built so much more rapidly, that it is economy to use full sheets anyway. It is very annoying to have crooked combs to deal with, and perfect combs cannot always be secured without the use of foundation (Fig. 91).

For extracting purposes, it is important, also, that the combs be built in wired frames (Fig. 92). It makes little difference to the comb honey producer whether his combs are wired or not after they are once built, because they are not subject to much strain. In the extractor, unwired combs are likely to be badly broken or ruined altogether. Fig. 93 shows a full sheet of foundation ready for the bees. Four horizontal wires are used

182   PRODUCTION OF EXTRACTED HONEY

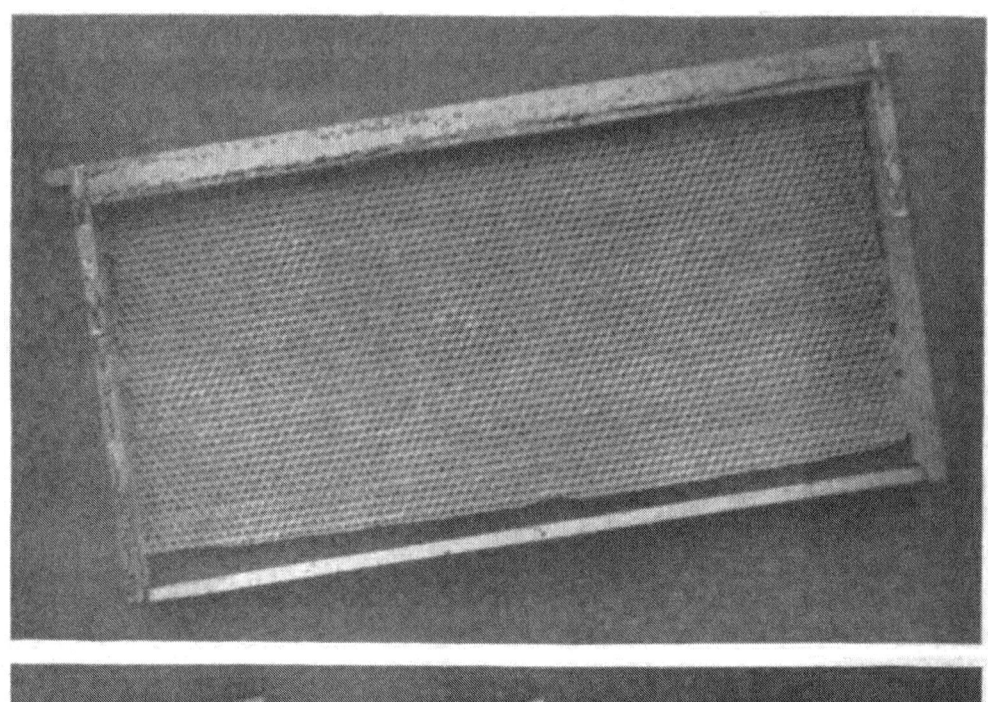

Fig. 93.—Hoffman frame with full sheet of foundation.
Fig. 94.—Development of combs from foundation.

in this frame. Fig. 94 shows how the bees make use of this foundation. The left hand frame contains a new sheet of foundation. The central figure shows the appearance when the bees are beginning to draw it out and the right hand figure shows a comb nearly drawn. Fig. 95 shows a good brood comb built from a full sheet of foundation on four wires.

The novice can seldom be made to see the importance of full sheets of foundation and wired frames. To save the extra expense of foundation, he will usually insist on using a narrow strip, with the result that his combs are not well built and are

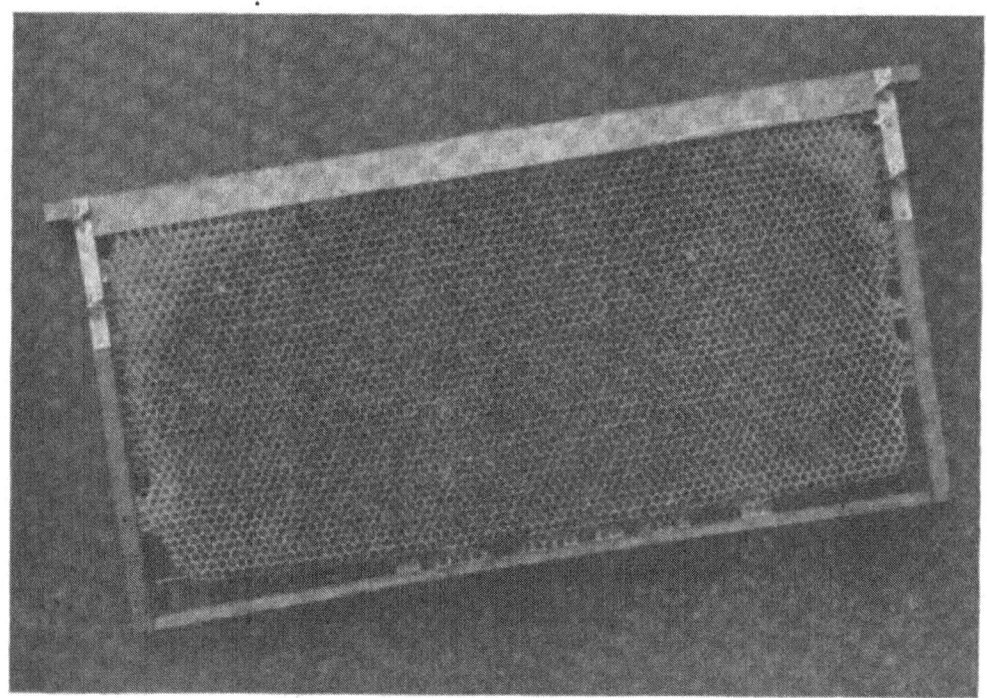

FIG. 95.—Comb built on wired frame with full sheet of foundation.

largely composed of drone cells. To avoid the trouble of wiring the frames, he will trust to the bees to build the combs strong enough, with the result that most of them will be broken the first time they are placed in the extractor. Experience is a good teacher, but here as elsewhere the tuition comes high. The use of proper precautions in the beginning would save much loss.

Strong Colonies Important.—What has been said elsewhere about the importance of having strong colonies at the beginning of the honey flow will also apply here. While medium colonies may store some surplus of extracted honey when they would not store in sections, it is only the strong colonies that pile up the profitable crops. No matter in what form one expects to market his crop, he must bend every energy to bring his colonies to the beginning of the honey flow in prime condition.

Putting on Supers.—As soon as colonies are crowding the brood chamber, it is time to give more room (Figs. 96 and 97). As the frames are the same as those occupied for the brood nest,

Fig. 96.—Strong colony for extracted honey production.
Fig. 97.—Colony that produced forty dollars worth of extracted honey in one season.

no difficulty will be found in getting the bees to occupy them, as encountered by the comb honey specialist. It is well to lift the hive up and place the extracting super underneath. By this means the bees will not be required to warm unoccupied space above the brood nest, and as fast as the honey comes in it will crowd the queen down, so that soon the upper hive will be full of honey and the queen and brood will be below. The objection to this plan is the accumulation of surplus pollen in extracting combs. If the empty super is placed on top without an excluder beneath there is a tendency for the queen to occupy the empty combs for egg laying, with the result that she will keep on going up as new supers are added, and more or less trouble will be necessary to separate the frames containing brood from those with honey only, at time of first extracting.

If empty supers are placed underneath, no harm will result, even though they be given some time in advance of when they are needed, and the extra room tends to keep down swarming.

Swarm Control.—It often happens that the extracted honey producer with his large hives has little difficulty from swarming, and need give the matter little special attention. The method of handling this matter most generally in use is known as the Demaree method. As soon as the brood nest is getting sufficiently crowded to require the addition of more room, the queen is hunted out and a frame of brood, preferably the one on which she is found, is lifted from the hive. An empty comb from the hive body used as a super is exchanged for it. The queen will then be on a frame of brood in a hive body of empty combs. A queen excluder is then placed on top of this new body and the old one already full of brood and honey is set on top of it in the usual place. The queen is now provided with an abundance of empty comb in which to lay. In fact her surroundings are similar to what they would be, had she recently come into possession of a new hive in company with a swarm. The colony will build up wonderfully in a short time, and not only will the desire to swarm be eliminated, but a tremendous working force will be present in the hive at the beginning of the honey flow. If additional room is provided as needed, further use of the excluder will hardly be necessary and it can be removed after two or three weeks.

In addition to the above advantages, the brood will be in the bottom of the hive, and the honey can be removed as fast as ripened and taken to the extracting room. While other methods of swarm control are practised to some extent in connection with extracted honey production, this plan is most generally used. It is also the simplest and surest in its results of any with which the author is familiar.

Use of Excluders.—There is a decided difference of opinion among bee-keepers as to the value of excluders. Aside from the above use, which is general at the beginning of the season to start the queen to laying in the lower story, many bee-keepers are of the opinion that there is little value in their use. Many are decided in the conviction that the use of excluders through the season results in a loss of honey. The author is of the opinion that they should not be used more than is necessary, although whether they actually result in smaller amount of honey being stored above them is apparently incapable of proof.

In comb honey production there is seldom if ever any occasion that justifies the use of a queen excluder. Occasionally some one will complain that the queen lays in the sections. This seldom happens anyway, and is of very rare occurrence, where full sheets of starter are used in the sections. The occurrence is so infrequent that it will neither justify the expense of excluders nor the inconvenience to the bees that their presence causes.

Ventilation.—In cool weather the entrance will furnish sufficient ventilation, but when the weather grows hot in midsummer more must be provided for best results. It is an easy matter to temporarily slip one hive body forward a half inch, the one above back a half inch, thus providing ventilation in every story from the bottom to the top of the hive. No rule can be laid down as to how much should be given. It will depend upon the weather and upon the honey flow. If plenty of honey is coming in so that there is no danger from robbers, much ventilation will be helpful in extremely hot weather. If no honey is coming in care must be used that the openings are not longer than the bees can guard safely.

Entrances the full width of the hive and at least an inch in depth are regarded as none too large for hot weather during the honey flow. Some lift the hive up an inch from the bottom board, and support it with blocks at the corners as described under comb honey.

The entrances should gradually be reduced as the season advances, the honey flow ceases, and the weather becomes cool in fall. A three-eighths-inch entrance is large enough for winter and even that is restricted to from four to six inches in width.

Ripening the Honey.—The practical bee-keeper will always provide a sufficient number of extracting combs, so that no honey need be extracted until it is fully ripened. A shortage of combs brings a temptation to extract too soon. Green or unripened honey should never be extracted. Some extensive honey producers are sometimes guilty of this practice. Not long since, the author visited an establishment where large quantities of honey are handled. A short time before a carload of extracted honey had been received from the West that had not been properly ripened. About one-third of this green honey was souring and working in the cans. Some of the cans had burst, and the whole thing was in such a condition as to demoralize any market where it happened to land. A few days longer on the hives, giving the bees time to evaporate it and ripen it fully, would have made a fine article which would have pleased the buyer, instead of causing him to curse the whole honey business. As a matter of course it was nearly a total loss to the producer. Why men will be so short sighted is hard to understand. The fact that they can sometimes sell the honey and leave the buyer to stand the loss leads them to risk it again.

Honey is seldom ready for extracting until the cells are nearly all sealed. Well-ripened honey can be kept for years without injury if properly cared for.

Removing Honey from the Hive.—Escapes are used to some extent in taking off extracted honey, as described under comb honey. It is a difficult matter to reach the bees in the sections and to get them out of the comb honey supers without escapes. Most bee men in taking off extracted honey open the hive and lift out a frame at a time and brush or shake the bees in front of the hive. The comb is then set in an empty hive body brought for the purpose. Full supers of frames are then set aside and covered until a load is ready to be taken to the extracting house. If there is an extracting room near at hand they are wheeled in, in a cart or wheelbarrow (Fig. 98), or if they must be taken some distance to the central plant, they are set in a wagon or automobile in which they are hauled home.

If escapes are used they must be put in place the day before

Fig. 98.—Wheelbarrow load of extracting supers.

the honey is to be taken off, which is often inconvenient, especially at outyards.

Extracting at Once.—The honey can never be extracted as easily as when first taken from the hives in warm weather. It sometimes becomes necessary to leave a part of the work to be done after the close of the season. If the weather is cool, a warm room will be necessary and even then honey that has stood in the honey house for several weeks will be thrown out with more or less difficulty. With a power-driven extractor it is possible to get the combs much cleaner than with the hand machine.

Most bee-keepers make a practice of extracting several times during the season, thus requiring less equipment and keeping honey from the different sources separate.

If one sells in a wholesale market, it is important to keep the light honey from clover and basswood separate from the dark, fall honey, such as buckwheat, golden rod, etc. It is better to extract after every flow as far as can be done, so as to keep the different kinds as nearly separate as possible.

If, on the other hand, the bee-keeper has a retail trade of his own and blends his product anyway, there is no special importance in keeping the honey separate, unless something might be brought in so poor in quality as to injure his crop. In many localities in the Southern States, there is a bitter weed that blooms in midsummer which secretes nectar from which honey that is too bitter to be eaten is stored. Where any plant of this kind is to be dealt with, it is important to remove all surplus from the hive as soon as it begins to bloom, to avoid having good honey mixed with it. A very little of this honey will spoil a whole crop, so that it cannot be sold to advantage.

Straining the Honey.—With the greatest care there will be bits of wax and other refuse thrown off in the extractor, which must be removed from the honey before it is ready for market. If deep settling tanks are used, this surplus matter will soon rise to the top, where it can be skimmed off, or the honey can be drawn from the bottom of the can where it is clear. By this plan there always remains a quantity of honey at the last that is not ready for market until it is strained.

Various contrivances are in use for the purpose of straining the honey as it goes into the settling tank. Thin cotton cloth is most often used as a strainer. A large surface is necessary to prevent the cloth from clogging, when it must be cleaned or a new one used in place of it. If the cloth alone is used, the weight of the honey will often result in pulling it loose at one side, when the whole of the contents will run into the receptacle below. A coarse screen of about one-fourth inch mesh is good to furnish a support for the cloth. If a large basket, which may be hung in the tank, is made of this coarse screen and lined with cheesecloth it makes a fairly satisfactory strainer. There is always more or less bother with clogged strainers, unless the basket is deep enough so that much of the refuse will come to the top rather than fall directly on the strainer.

Alexander Strainer.—The Alexander strainer is made of fine wire screen, and is about the size and shape of a large bucket with bail. This pail is hung in the tank or other receptacle, in which the honey is stored and the honey run into it as extracted. The bottom and all side surface permitting the passage of honey, it does not clog readily and it is strong enough to sustain the weight of a full pail of honey. All sediment is caught and held. The strainer is easily cleaned with hot water after the refuse is dumped out.

Second-Hand Containers Not Desirable.—So much honey goes to market in the square sixty-pound cans that there is always an accumulation of them in all the large centers. These are offered for sale at a very low price. So little is to be saved by the use of these second-hand containers that the bee-keeper can hardly afford to buy them. If they are rusted inside, the quality of the honey will be injured, and if otherwise perfect there is some danger of spreading disease by their use.

As mentioned elsewhere the principal bee diseases are spread from hive to hive in the honey. Second-hand containers brought to the apiary are more or less daubed with the honey with which they have previously been filled. This honey attracts the workers, and if it came from a diseased colony there is great danger in bringing it into the apiary. Disease is thus spread to considerable extent. The author has had his attention officially called to this source of disease so many times that he is inclined to favor restrictions on the use of containers for honey a second time, unless it be in the same apiary where filled at first.

If the honey is put up in bright new cans a better impression is made on the buyer than if received in cans that are rusty and stained. Occasionally a buyer makes serious complaint if honey is received in such cans.

Liquefying Candied Honey.—After extracted honey has stood for a time, it will usually candy. If it goes to market in the sixty-pound cans in which it is stored, the producer will have no occasion to liquefy it, as it will stand the journey with less risk in this condition. If through any accident a can should be damaged, there will be no leakage, as would be the case if the honey was shipped in a liquid condition.

If the honey is to be placed in small packages for the retail trade, it will be necessary to heat the honey sufficiently to restore the liquid form. Great care is necessary not to overheat the honey, as to do so will greatly injure the flavor and consequently the value of the product.

Various plans of accomplishing this result have been devised. In large establishments a system of hot water pipes is sometimes used. The caps are removed from the cans, and they are set up-side-down on pipes. As fast as the honey melts, it runs out into a container below.

Large tanks are also used which are filled with hot water around the honey cans. This water is kept at a temperature of about 150° for a sufficient time to liquefy the honey in all the cans.

A simple and very satisfactory plan is illustrated by Fig. 99. This plan utilizes an ordinary cheap feed cooker such as can be purchased in the market for about twelve dollars. There is just room for eight sixty-pound cans in the square tank. Instead of using hot water, a crate of wood is made to hold the cans about two inches off the bottom and water is allowed to come just to the bottom of the cans. A lid shuts down, as will be seen in the picture, and a very light fire is started in the fire box underneath. As the water is heated steam is generated, and the cans are warmed by steam instead of having the hot water in contact with them. A small hole in the top of the lid provides a place for a thermometer, which indicates the temperature. One great advantage in this heater is that if, by chance, it becomes too hot the lifting of the lid permits the escape of the steam and cooling of the interior instantly. If the water system gets too hot, it is difficult to cool it quick enough to avoid injury to the honey. Several hours will be required to liquefy the contents of the

Fig. 99.—Utilizing feed cooker for liquefying candied honey by steam.

cans by this system, but the amount of fuel required is so small as to be a very insignificant matter.

Bottling.—If the honey is sold through retail stores a portion of it is likely to stand on the shelves for some time after it reaches the store. In this case the contents of many of the bottles will candy again in time. Sometimes a trade is developed that comes to demand a certain amount of this candied honey or will take a jar of candied honey and liquefy it by setting in a pan of warm water. However, in most localities, the bee-keeper will be required to take back honey that has candied and replace it with honey in the liquid state. It is an easy matter to restore the honey in jars in a few minutes by setting them in a shallow tank of hot water that just comes up around the necks of the bottles.

If the honey is kept at a temperature of about 120° for several hours before bottling, and then sealed while still warm, several weeks and sometimes months will often elapse before it will candy again.

Some bee-keepers make a practice of restoring honey that has candied in small glass jars by placing them in solar wax extractors, where they are exposed directly to the heat of the sun. This plan seems to be very satisfactory for small quantities, as the sun's rays supply about the right conditions for best results.

Retailing Candied Honey.—Some honeys have a much greater tendency to candy than others. Western alfalfa honey candies very quickly and becomes quite hard. Some honey will only candy far enough to become waxy and sticky. Unless it becomes hard enough so that it is no longer sticky, there is little opportunity to develop a special trade for candied honey in small packages. Several kinds of pasteboard or paper packages holding small quantities of this honey are in use. The paper bucket commonly used for retailing oysters is perhaps the most commonly used. When the honey shows signs of granulation, but will still run, it is drawn into these packages and set in a cold place. Frequent changes of temperature hasten granulation and a room where it is first warm, and then freezing, will be the best for honey which it is desired to granulate. When the honey is sufficiently hard, it is placed on the market. Unless subjected to quite a warm temperature it will remain in the granulated condition for an indefinite period.

As yet there is no general market for granulated honey in these small packages. Every bee-keeper who wishes to handle honey in this way must develop his own trade. It would seem that a nice trade might be gradually developed for small cubes of this candy to sell at a nickel through the retail candy trade. Once people came to know the product they would buy it freely, if it were available in a five-cent package.

Once the public is educated to understand that only honey of the best quality can be marketed in this form, the bee-keeper will find a ready market for candied honey.

BULK OR CHUNK HONEY

In many localities in the Southern States there is a demand for bulk or chunk honey. The general principles of producing extracted honey will also apply to bulk honey. The foundation in the supers need not always be of full sheets, nor should they be wired. An empty comb or two in each super will be helpful in getting the bees into the new super promptly. When the combs are finished they are cut from the frames and new foundation is put in for future use. Bulk honey can be produced cheaper than section honey, but not as cheaply as extracted honey, as the combs must be built new each time the crop is removed. With extracted honey the combs can be used again and again, which makes larger production possible under ordinary conditions.

QUESTIONS

1. Note the difference between strained honey and extracted honey.
2. Discuss extractors and other equipment for the production of extracted honey.
3. What kind of frame is most satisfactory?
4. What things are essential in a honey house?
5. Describe different kinds of honey houses with advantages of each.
6. How should doors and windows be screened and why?
7. Why are full sheets of foundation in wired frames desirable?
8. How and when should supers be added?
9. Describe the Demaree method of swarm control.
10. Discuss queen excluders.
11. How much ventilation is desirable and what size entrances should be used?
12. When should the honey be taken from the hive?
13. Discuss extracting and straining of honey.
14. How should honey be stored?
15. Discuss candied honey.

CHAPTER XI

WAX A BY-PRODUCT OF THE APIARY

ALTHOUGH honey is the principal product, considerable wax is produced in every well-regulated apiary. Although bringing the highest price of anything the bee-keeper has to sell, the possibilities of this special output are too often overlooked because much of it is gathered in small quantity in scraping sections, cleaning burr combs from the tops of frames and scraps of combs that accumulate about the bee yard and honey house. If the bee-keeper who has not carefully saved these odd bits of comb will provide a bucket or other receptacle which is always kept at hand in which to place all scrapings and bits of wax he will be surprised to see what a quantity will accumulate during the season. In addition the apiary and equipment will be much cleaner as a result. It is very annoying to the housewife to have someone coming into the house with bits of wax clinging to his heels to be left on the rugs or carpet, as will frequently be the case where such refuse is dropped on the ground about the bee-hives.

Old combs that are to be discarded and cappings which are present in quantity are usually saved, as they should be, but unless some care is used they are likely to be destroyed by the wax moths during the warm weather. It is a good plan, no matter what method of wax rendering may be adopted, to throw all such material into a solar extractor at once. In this way it will be melted so thoroughly that there is little trouble with moths, even though it is not separated sufficiently to avoid the necessity of rendering.

Production of Wax.—When the bees are feeding heavily, as during a good honey flow, wax is secreted as a direct result of the quantities of food consumed. After a colony has swarmed in warm weather large numbers of bees will cluster together apparently for the purpose of secreting wax and with it building the new combs which will be necessary to store the food supply and rear the brood of the colony. The wax pockets are eight in number for each worker-bee. They are located on the under side of the abdomen, four on each side. By watching the bees at times such as above mentioned, the little wax scales can be seen protruding between the segments of the abdomen. The author is not sufficiently gifted to describe the wonderful manner in which they utilize these minute scales and the way they manipulate them to form the perfect combs which are so essential to the welfare of the colony. No description will satisfy the enthusiastic bee-keeper who must see it all for himself. By providing an observation hive at the proper season many interesting operations may be seen. The worker may be seen to take the wax scale in her jaws and to knead it, apparently, after which it is added to the partially built comb which her predecessors have started. But a moment is thus occupied when she moves away and her place is taken by another who also adds her portion. The work is done very much as though men in building a wall each brought a single brick and put it in place and went away. Yet in spite of the apparent hit and miss method of building, there is no more wonderful or more perfect structure than the combs of the honey-bee.

Wax melts at a low temperature, as many a bee-keeper has learned to his cost when brood combs have been left exposed to the hot sun on a summer day. At times the heat is sufficient to melt the combs within the hives, especially when they are surrounded by high board fences, dense undergrowth, or other obstruction that prevents a breeze from reaching them, or if the hives are not well ventilated.

The young bees do most of the work of comb building, as the ability to secrete wax declines with advancing age. In case of necessity old bees will build combs, although apparently they secrete wax less readily and in smaller quantities than the younger ones.

PRODUCTION OF WAX

Color.—There is a great variation in the color of wax, depending upon the source of the food supply of the bees at the time of comb building. As a rule newly built comb is light in color, gradually growing darker with use. The brood combs shortly become quite dark, and in time almost black, due to the stains of constant travel as well as refuse from the growing larvæ and the cocoons which are left behind when they emerge from the cells. When old combs are melted, so many of these cocoons often remain that they will retain the exact shape of the original cell.

Size and Shape.—The difference in size and shape between the cells prepared for various purposes, as for the rearing of queens, is so striking as to attract instant attention on looking within the hive. Much has been written in admiration of the mathematical precision with which the bees are able to occupy all the available space by building a six-sided cell, the bottom of each of which was opposite the bottom of one-third of each of three others. By building in this way the maximum of both capacity and strength is secured with no lost space.

If the bees build according to their own plans the combs are usually about an inch in thickness with cells of equal depth on each side. If built within frames in a hive they may be thicker or thinner, depending upon the spacing of the combs. Extracted honey producers often space their frames so as to secure thicker combs to make the work of uncapping easy. The distance between the combs is from three-eighths of an inch to seven-sixteenths of an inch, depending upon circumstances. The bees require about three-eighths of an inch at least in order to move about easily. Combs are usually placed about an inch and a half from center to center.

The worker cells are the smallest and we resort to the use of foundation to insure that the cells will mostly be built of this size, as mentioned elsewhere. According to most writers each worker cell is about one-fifth of an inch in diameter, and the drone cells are somewhat larger. The queen cells are built especially for the particular purpose of rearing queens and are built only as needed and frequently torn down when no longer of immediate use. The regular comb built permanently is all of the six-sided shape and of the two sizes. The larger cells such as are used for rearing drones serve equally well for honey storage.

Uses of Wax.—For many centuries beeswax has been known as a commercial commodity. So valuable was it in ancient times that taxes were at times paid in wax and a tribute of wax was levied by victorious kings on the unfortunate inhabitants of the country which they had overrun. Many references to this product are to be found in ancient writings both sacred and secular. Rents and other obligations were paid in beeswax to such an extent as to indicate the demand must have greatly exceeded the supply. Before the invention of paper, wax tablets were used for the purpose of making temporary records, for correspondence, etc.

Wax candles have long been used for various ceremonial purposes in the churches, and this custom has survived the centuries and still offers a market for quantities of wax, for some churches still use candles made of beeswax for this purpose.

Many delicate objects are moulded of wax, as fruits and flowers, that are so natural in appearance as to perfectly deceive the casual observer. Figures and models of various kinds are also made of this material, as it is very plastic and responds to the most delicate touch of the artist.

Tailors make use of pure beeswax in many cases for sewing wax, shoemakers and harnessmakers also make use of it, either pure or mixed with other materials for waxing their threads.

It is a common ingredient of varnish and furniture polish, lithographic inks, various cements, waterproofing materials, and in many remedies and other commodities handled by the drug trade.

Comb Foundation.—The bee-keeper has of late years come to be his own best customer. Since the invention of the mills that make comb foundation possible, extensive use has been made of it among the bee-men themselves.

Next to the movable frame hive, comb foundation has perhaps made possible the greatest advance in bee culture. Without the use of foundation it is a very difficult matter to get straight combs or to prevent the bees from building crosswise or otherwise than according to the bee-keeper's wishes. With the use of foundation the possibilities of honey production are multiplied and no practical honey producer would think of doing without it.

Only pure beeswax should be used in foundation as otherwise the sale of honey in combs built on it would be a violation of the pure food laws. Fortunately little if any adulteration of comb foundation is practised, the manufacturers being very careful to test all wax used for the purpose and the bee-keeper can buy from any of the well-known manufacturers with confidence.

The wax is melted in the factory and wound in long sheets which are run through mills bearing the impression of the size and shape of the worker cells. As the foundation is printed it is cut in strips of convenient length and these are wrapped in thin paper to prevent sticking together when warm. The papered strips are then packed in paper boxes in such quantities as the needs of the market demand. Use of foundation is considered in the chapters relating to comb and extracted honey.

Substitutes for Beeswax.—Various mineral and vegetable waxes have taken the place of beeswax in various commercial uses. These waxes can be produced much cheaper and answer fully as well for many purposes. Paraffin, ceresin and several others are well-known commercial products. Substitutes for wax made into foundation will not be accepted by the bees.

Adulteration of Wax.—Dealers who buy beeswax must exercise constant vigilance to avoid being imposed upon by an adulterated product. As the adulterations can be purchased at prices much below that of beeswax, dishonest men see possibilities of great profit if they can sell their dishonest product. Various tests have been discovered for detecting the adulterations until it is now very difficult indeed to get adulterated wax to market without detection.

Paraffin, ceresin and sometimes tallow are common adulterants of wax. Wax is so commonly adulterated that when it reaches the market it will be subject to very careful examination and any fraud is likely to be discovered.

WAX RENDERING

Commercial establishments which deal in wax are so well prepared to render the wax at a low price that many bee-keepers ship all combs and refuse containing wax to some of these establishments at the end of the season. Either the bee-keeper pays cash for rendering the wax and has it worked into comb foundation for future use, or he sells the wax for cash and is charged a small fee for rendering. Where the bee-keeper has but a small amount of material this is frequently the most satisfactory way of disposing of it, as he avoids a very mussy job at best and his time may often be otherwise employed more profitably.

The Solar Extractor.—The solar wax extractor is made by placing a glass a few inches above a sheet of metal which is tilted enough to allow the melted wax to run off and depending upon the heat of the sun to melt the combs. New and tender combs or cappings will be pretty well rendered in this manner but old combs will not be well separated. In any case a solar extractor is a valuable item of equipment in an apiary for bits of comb can be thrown into it as collected and thus be saved. Old combs may be melted to prevent damage by moths. Considerable quantities of wax will accumulate in the wax box at the bottom and this will save handling again later. The whitest and best wax will be secured in this way. It will nearly always pay to render the refuse from a solar extractor in a wax press as otherwise much of the wax is wasted.

Boiling in a Clothes Boiler.—There are a number of crude methods by which bee-keepers with but a small amount of wax have long extracted it. One of these is to boil the combs in a

THE WAX PRESS

wash boiler and to skim the wax from the surface of the water. Sometimes the combs are placed in a burlap bag and thrown in the boiling water. Sticks are used to punch the bag and to stir it about in the hot water. While a certain amount of wax will be secured in this manner it is very wasteful and from one-fourth to one-half of the wax will be lost unless the refuse is rendered again by some plan.

Small bits of comb are often placed in a pan in the oven. The pan is partly filled with water and the hot wax dipped off or the combs are laid on a screen through which the wax will run while the waste will remain on the screen. The wax is sometimes left to harden in the pan and the cake lifted out when cool.

While rendering by some of these crude methods is better than wasting the wax the amount wasted will shortly pay for a good press.

The Wax Press.—No satisfactory way to get all the wax has been found without the use of some kind of press. Some may think that they are getting all the wax because the slumgum or refuse is apparently free from it, but the chances are that when rendered with a good outfit this slumgum would produce from fifteen to twenty-five per cent more wax.

A man who understands mechanics and is handy with tools can readily construct a wax press, though there are good ones to be had in the market. The principal requirements are great pressure applied when the mass is hot, and that there be plenty of water mixed with the melted combs to insure that the wax will run freely. Many of the outfits in use have some provision for the use of steam to keep the whole thing hot when the pressure is applied. It has been found of late that the press need not be heated if the work is done when the weather is warm or in a warm room, providing that the material is boiling hot when dipped into the press.

Many different plans for making presses for this purpose have been described in the bee journals but the essential requirements are the same. Some are round and some square but with pressure properly applied and the material of the right temperature almost any of them will get the wax.

The Hatch Press.—The most popular press seems to be a modification of the Hatch press and is usually called by that name. It is a simple implement as good things usually are (Fig. 100). There is a strong frame to give sufficient strength to the iron screw by means of which the pressure is applied. The form is of metal and round in shape. There is a round opening for drawing off the liquids. This may be closed with a cork when desired to prevent the escape of the hot water when pressing.

Fig. 100.—Hatch wax press.

How the Wax is Rendered.—In addition to the press will be needed a boiler in which to melt the combs and some sheets of burlap. The boiler should be partly filled with water and placed over a hot fire. The combs to be rendered are thrown into the water and after it begins to get hot are stirred freely. As the mass is melted additional combs can be added. Plenty of water should be used to insure best results. The mass should be heated thoroughly and all wax should be melted fully but care should be used not to overdo it and scorch the wax, as might easily happen if too many combs are boiled for the quantity of water. The burlap should be laid in the press and the opening corked up and the tank filled with hot water in order to have the press hot when the melted combs are poured into it. The water can then be drawn off and set aside for use if needed. The burlap is then used to line the form and a quantity of the mass from the boiler is poured into it. A large dipper is a good thing for this purpose. The wax should not be strained before pressing but water and all should go into the press. The ends of the burlap are now turned over the mass so that it cannot escape excepting as strained through the cloth. The cleated follower is then placed on top of the cheese and the whole thing is set in place under the screw. The screw is turned slowly down as long as liquids can be squeezed from the " cheese." The water and hot wax will run off together through the opening in the side of the can. When no more wax is coming the screw can be released and the " cheese " doubled up and given another pressing.

The " cheese " is then removed and thrown to one side and another lot is pressed until all the combs have been rendered. Usually it will be necessary to break up the " cheeses " that remain and boil them again, and again press them as in the beginning to get all the wax. This second rendering will usually bring a surprising amount of wax from material that appears to be entirely free from it. Tests from various samples have shown from ten to fifteen per cent of wax still in the slumgum.

The Massachusetts Agricultural College has opened a wax rendering station for the benefit of the bee-keepers of that State. Bee-keepers are instructed to use a barrel for storage purposes and as combs and bits of wax are thrown into the barrel to tamp it down tight and when full to ship to the station for rendering. If smaller quantities are to be shipped some smaller container can be used. This station is proving to be very popular with the bee-keepers of that State and large quantities of comb are being shipped to the station.

The Steam Press.—At one time steam wax presses were in common use but they are generally being replaced by the method previously described. The steam press is heated by steam generated from water in the bottom of the can. It will be necessary to set the press on a hot stove or to make some provision for heating the water. Above the water is a basket to hold the combs on which pressure is applied by means of a screw. The melting wax falls into the water below and runs out the overflow spout (Fig. 101).

Boiler Press.—There are different kinds of hot water presses in use but in general they may be said to consist of a strong can in which is contained a heavily bound basket. A bar across the center supports the screw by means of which the pressure is applied. In this kind of extractor the water comes up around the melted combs which are under pressure and the boiling and pressing are carried on at the same time. When the wax is all out sufficient water is supplied to carry it off through the tube near the top, while the small amount of refuse straining through the cloth settles to the bottom of the can. This plan gives good results if carefully done but there is some difficulty in getting all the wax out of the can.

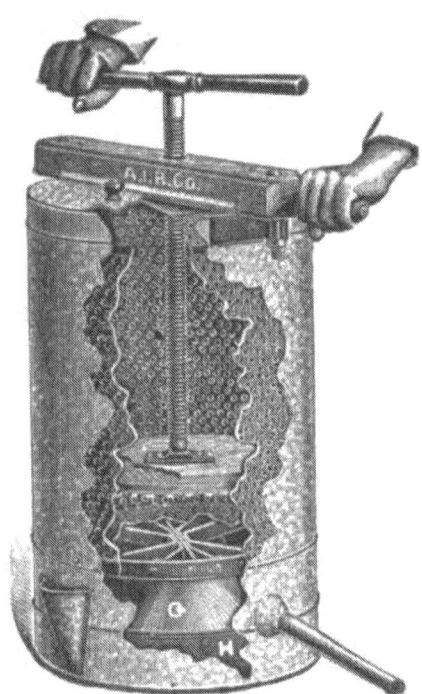

Fig. 101.—Steam wax press.

There is no trouble about the mass cooling while under pressure and the operation can be repeated as often as desired by simply loosening the screw and saturating the cheese with water again.

Bleaching Wax.—Every time the wax is melted the tendency is to a lighter color and the exposure to the sun in the solar extractor also tends to whiten it. Although sulfuric acid is sometimes used for clarifying, there is seldom any occasion for the bee-keeper to bother about bleaching further than to re-melt any cakes of wax that are very dark and to remove as much impurity as possible. The difference in price that will be received will hardly pay for the extra trouble, however.

Cooling the Wax.—Utensils into which the hot wax is poured for cooling should first be dipped into cold water or greased to prevent the wax from sticking. Then care should be used to prevent the wax from cooling too rapidly or the cakes will crack.

QUESTIONS

1. How is wax produced?
2. Of what use is it to the bee-keeper?
3. Discuss the various commercial uses of beeswax.
4. How is comb foundation made?
5. Why is beeswax often adulterated?
6. Discuss the different methods of rendering beeswax.

CHAPTER XII

DISEASES AND ENEMIES OF BEES

As a country grows older new vicissitudes beset almost any line of business, and bee-keeping is no exception. In many sections of the United States brood diseases have not as yet appeared, and in many others the bee-keepers are having their first experience in combating them. However, it is only a matter of time until bee-keepers can expect to be compelled to deal with foul brood no matter where they live. It accordingly will pay the business bee-keeper to inform himself as fully as possible concerning bee diseases, even though there be none at present in his vicinity.

Expert bee-keepers are frequently all but ruined by the appearance of foul brood in their apiaries. With a thorough knowledge of the accepted methods of dealing with disease the experience need not be so costly, for by prompt action the danger can be largely avoided.

The census of 1901 showed a decrease of 16.1 per cent of the total number of colonies of bees in the United States as a whole. The wide-spread presence of disease is no doubt largely responsible for this condition. With an increasing population and a decreasing number of bee-keepers, it would look as if the business of honey production should offer a good field of operations. While the small bee-keepers with a few colonies on the farms are rapidly being removed, specialists are increasing in number. This is as it should be, for to-day is the great age of specialists and the business which is not worthy of development as a specialty offers little inducement to the active man.

While there are still many puzzling things that manifest themselves in connection with foul brood, the essentials necessary to the control of either form are pretty well understood and practical men who are on their guard find it possible to withstand the onslaught without great losses. It usually happens, however, that disease has gained considerable headway in the apiary before its owner is aware of the nature of the difficulty. Especially is this true when disease puts in an appearance for the first time in a locality that has been free from it. The journals frequently recount the experience of some unfortunate who has suffered heavy losses in this manner. The writer, in the capacity of State inspector of apiaries, sees such cases very frequently. Instances have come to his attention where the losses amounted to many thousands of dollars, whereas had the owner realized the nature of the trouble on its first appearance it could have been checked without difficulty.

AMERICAN FOUL BROOD

Much confusion has resulted in the similarity of names of the two common diseases. It is unfortunate that some entirely different name was not applied to one or the other. While European foul brood has long been known, in some localities, under the name of "black brood," the name was not appropriate and it has given way to the accepted title of European foul brood. There is a decided difference in the appearance and in the action of the two forms, so much so that there need be little difficulty in recognizing the difference in advanced stages. In early stages it is sometimes a little difficult to determine which form one may have to deal with, and in that case it is well to cut out a piece of comb containing the dead larvæ and, wrapping it securely, send it to the Bureau of Entomology of the United States Department of Agriculture. By means of a microscopic examination they can readily determine the nature of the difficulty.

American foul brood has long been present in this country and when we hear the term "foul brood" we naturally infer that American foul brood is meant (Figs. 102 and 103). It is also called "ropy foul brood" because of the peculiar ropy characteristic of the dead tissue at a certain stage. The larvæ are usually attacked at about the time the cells are capped and most

Fig. 102.—Brood comb from colony affected with American foul brood. Three adult wax moths can be seen on the comb and tunnels of the larvæ are beginning to appear.

Fig. 103.—Work of wax moths in colony affected by American foul brood. The sunken cappings are typical of American foul brood.

of the cells containing dead larvæ are capped. When the larva dies it turns a chocolate or brown color and in advanced stages of decay becomes darker. The cappings become sunken, and frequently the cappings are perforated by small holes. The most common test for this disease is to insert a toothpick or timothy straw into the dead tissue and slowly withdraw it. The decaying matter stretches out like thick molasses, sometimes for an inch or two before breaking. After the dead larva has become fully dried it forms a dried scale on the lower side of the cell. This scale adheres tightly to the cell and can be readily observed by holding the comb in front of the eyes at such an angle that the light falls into the bottom of the cell and illuminates the lower side wall.

There is also a very characteristic odor clinging to the combs containing a badly infected case of foul brood of the American form. It is commonly spoken of as a glue-pot odor but that hardly describes it. It is, however, a characteristic of the disease that can readily be recognized. Queen and drone larvæ are seldom affected by American foul brood, while the other form attacks both queen and drone larvæ at the same stage as worker larvæ are affected. Cases are reported where there is a decided odor with European foul brood, but the writer does not remember ever having seen a case in all the hundreds of apiaries visited. The ropy condition of the dead matter together with the odor is usually considered as positive evidence of American foul brood.

When this disease is present the death of a portion of the brood gradually decimates the colony until it becomes so weak that it can no longer defend its stores and it is likely to be robbed out and the honey carried to other colonies. The disease is thus spread far and wide. The writer has seen cases where after the death of the colony from foul brood the hive was turned over and exposed to the bees by the owner, who was ignorant of the real cause of the trouble. The disease was thus needlessly carried into every colony of large apiaries.

The disease is caused by a bacillus technically known as

Bacillus larvæ. These microörganisms are so extremely minute as to require a high power microscope to enable one to find them. The germs or their spores seem to be carried from hive to hive only in the honey. In treating American foul brood it accordingly becomes exceedingly important to rid the colony of every vestige of the diseased honey. While the honey may carry the germs of foul brood which are fatal to young bees, it is not in the least injured thereby for human consumption.

It is important that this point be fully understood or otherwise any method of treatment is likely to be unavailing. On one occasion an inspector was called to examine the bees in a neighborhood where foul brood was known to be present. At one farmhouse he was told by the housewife that they no longer had any bees but some empty hives. On investigation he found that the bees had died during the winter from American foul brood. It was still early spring and the honey had not yet been found by the bees of the neighborhood. He explained carefully to the owner the method of treatment and thought that he fully understood it. The next day a man was sent back to ascertain whether instructions had been properly followed, only to find that he had carefully disinfected the hive by burning it out, but had left the honey lying on the ground where it was even more likely to be found by visiting bees than had it been left in the hive. In this case a large apiary near at hand was saved from infection by the fortunate visit of the inspector.

Bees weakened by disease are very likely to die in winter. In such cases the old combs should in no case be used again, but the wax should be rendered and the hive carefully disinfected before being put in service. Colonies thus weakened are also very likely to fall an easy prey to the wax moth, and it frequently happens that colonies which are charged to the ravages of the moth are really victims of foul brood. It is frequently recommended that honey from diseased colonies be boiled and fed back again to the bees. While this may be safe if carefully done, it is much safer to feed sugar syrup if it becomes necessary to feed anything. It is regarded as unsafe to feed the honey from hives infected with this disease, as high temperature for considerable length of time is necessary to insure death of all spores. In the hands of a novice it frequently happens that the boiling is not sufficiently thorough and healthy bees are thus infected.

Treatment of American Foul Brood.—This disease is rather slow in its progress, but very sure, and once a colony becomes infected its final death is certain, unless the bees are removed to a clean hive and the infected brood destroyed. In the hands of the average bee-keeper the shaking treatment, commonly called the McEvoy treatment, is best. McEvoy, who was for a time inspector for Ontario, was very successful in treating foul brood and he it was who probably first brought successful methods of treatment prominently before the public. However, the essentials of this method were described in Europe many years before the birth of McEvoy, and Quinby had also long made use of shaking for the cure of foul brood in this country.

The first essential is to remove the bees entirely from the source of the disease, and they should accordingly be placed in a clean hive on the old stand and the old combs, brood, and honey all removed. McEvoy allowed them to build new combs for four days, thus insuring that all honey carried with them would be used, and then again shook them into another clean hive and destroyed the combs that they had built in the meantime. The second shaking is not always necessary. By using good judgment the bee-keeper can usually tell when conditions are such that a second shaking will be necessary.

The *instructions* given from the office of the Iowa Inspector are as follows:

> In the evening after the bees have stopped flying, brush or shake all the bees into a clean hive containing foundation starters. Bury or burn the old combs at once, not the next day. Take great care that no honey, not even the smallest drop, be exposed to the bees, or the disease may be carried back or exposed to healthy colonies.

This is essentially the instruction given for years past by various State officials charged with enforcement of foul brood laws. It is repeated here simply to show that the essentials can be stated in a few words.

Modification of Method.—If the bee-keeper does not give the second shaking at the end of four days he should watch very carefully to see that the disease does not again appear. There are a number of modifications of this method of treatment, each of which has advantages apparent to those who follow it. Thomas Chantry inserts a dry extracting comb in the center of the hive on which the bees are shaken and about twenty-four hours later very carefully removes this comb. In the meantime the bees will have used the empty comb to deposit the honey that they may have carried with them. This is much to be preferred to the second shaking as it saves a heavy loss in wax secretion and consequent tax on the bees which are badly used at best. Edward G. Brown, of Iowa, who is a large honey producer, has used this method successfully for a number of years and recommends it as very satisfactory if carefully done.

D. E. Lhommedieu, another Iowa bee-keeper of long experience, shakes the bees into a clean hive and leaves them for four days or until he is sure that all old honey carried with them has been consumed. He then takes combs of brood and honey from healthy colonies and places them in a clean hive and puts this on the stand where the diseased colony has been. Feeling that the bees have rid themselves of the infection, he proceeds to shake the bees into the new hive containing the brood and they are thus saved the heavy tax of building up from the beginning.

The object is to rid the bees of every trace of the diseased honey before the new brood appears in the hive and any method that will accomplish this result is likely to succeed.

When a number of colonies are to be shaken, it is well to replace the frames of brood in the old hives and to pile one above another on top of some diseased colony which may be reserved for treatment for a few days, until the healthy brood is hatched, and thus save what healthy brood there is in all the hives. This plan has been carried out very successfully in some apiaries.

One of the best methods of treatment is to remove the queen very carefully, disturbing the bees as little as possible. The hive should then be tightly closed with the exception of a bee escape, which will permit the bees to go out but give none a chance to return. Take a frame of healthy brood from some other colony and place in a clean hive. Fill the remainder of the hive with full sheets of foundation or empty combs and place it where the colony has stood. The queen may then be placed on the frame of brood and the new hive left with the entrance somewhat smaller than usual. Turn the hive containing the diseased colony around so that the escape will be near the entrance of the new hive. The bees leaving the hive go to the fields with their honey sacs empty and returning enter the clean hive. As fast as the brood hatches in the old hive the bees will leave only to find no way of return and enter the clean hive in which the old queen is at work as usual. This method has the advantage of saving the colony without loss of brood or checking the laying of the queen. If properly done this is perhaps the best method of dealing with American foul brood. Some bee-keepers advocate setting the diseased colony on top of the clean hive with the bee-escape board underneath and the old queen left in the brood chamber. By this method the bees will rear a young queen from the brood in the frame given them in the clean hive while the old queen continues to lay in the diseased chamber above until she is finally deserted by the workers.

Late-Season Cases.—When a case of foul brood is found in fall after the honey flow is over, it is seldom advisable to attempt to winter the colony. In general it may be said that treatment is not likely to be successful, excepting when there is some honey coming from the fields or will be later in the season. If cases are treated ahead of the honey flow, the lack of a flow can be met by heavy feeding to stimulate the building of combs. If the bees get well started in this way they will recover nicely during the honey flow that follows later on. After the flow is over in the fall it would cost more than they are worth to feed a sufficient amount of stores to build them up ready for winter. To winter a colony with the idea of treating in the spring will require in the neighborhood of twenty-five pounds of honey, and there is always the danger that they may die during the winter or early spring. In this case there is not only the total loss of the bees and the honey that they have consumed, but the added danger that bees from other colonies may get at the stores and rob them out on some warm day before the hives have been looked after, and the disease be further spread. If the colony is strong enough to have a fair chance of wintering it is possible to save honey and wax to the value of from two to four dollars, and this is more than a diseased colony is worth at this season of the year. The hive may be saved and prepared for use again by proper disinfection.

Late in the evening after the bees have stopped flying, the entrance should be tightly closed to prevent the escape of any bees. The hive should then be removed to some tight building or cellar and the bees killed with sulphur. All honey fit for use can be removed, but care should be taken that not a drop ever gets back to live bees. The combs can be melted up and the wax saved. Honey not fit for the table can be made into vinegar. The hive, including both top and bottom, should be thoroughly disinfected before using again, and if the frames are to be used again they should be boiled. Any honey that is fed to bees should be diluted with water and boiled for half an hour or until the scum is thoroughly cooked.

Disinfecting.—For disinfecting hive parts a painter's torch is very good. Some paint the inside of the hives with kerosene and then pile one above another and set fire to them and smother the fire as soon as the interior is scorched.

EUROPEAN FOUL BROOD

The cause of European foul brood is supposed to be *Bacillus pluton,* a microörganism similar to those responsible for such diseases as diphtheria, typhoid fever, etc., in human beings.

Authorities are not agreed as to the method of spread of this disease. That it is not spread altogether in the honey as is American foul brood is evidenced by the fact that strong colonies with vigorous young Italian queens frequently clean out the infection and that it does not reappear in the hive. In the case of a colony affected with American foul brood the final death of the colony seems assured unless the last trace of the diseased brood and honey is removed. While some authorities for a time insisted on shaking in treating for European foul brood the same

Fig. 104.—Thirteen colonies left of one hundred five as the result of European foul brood for eight months.

as for American, shaking is no longer advised in the treatment of this disease. It is now well known that the destruction of combs and honey is not necessary in dealing with European foul brood.

One striking peculiarity of this disease soon becomes apparent to an inspector; when it appears in a malignant form it is usually to be found in every colony in a yard within a short period of time. While American foul brood may be present in a yard for months without spreading, European foul brood frequently, though not always, spreads very rapidly and appears in all colonies very quickly. Cases have come under the writer's observation, where no disease had been present in a locality, European foul brood suddenly appeared in nearly every colony of several large apiaries situated near together. At times it seems very mild and will even disappear of itself. At other times large numbers of bees will die in a very short period of time. The illustration (Fig. 104) shows a case where but thirteen colonies remained of one hundred and five in eight months. The disease was not known to be present until two weeks after the bees were taken from the cellar in spring, only about six weeks before the picture was taken. As the winter loss was unusually heavy it is presumed that the disease was present when the bees went into winter quarters.

American and European foul brood, it would seem, can be compared to smallpox and typhoid fever in the human race. American foul brood, like typhoid fever, requires a common source of infection, in the case of the bee disease the honey, in the case of the human ailment milk, water, etc. European foul brood seems to spread among bees as readily as malignant smallpox among the human race, actual contact apparently not being necessary to the spread of either. However, until recently little was known about European foul brood and it is entirely probable that later discoveries will add much to our knowledge of the disease.

Appearance of Affected Larvæ.—European foul brood attacks the larvæ at a much earlier stage than does American foul brood and but a small part of the diseased brood is ever capped (Fig. 105). In bad cases large numbers of the larvæ will be found to be dead and misshapen while still white as shown in the plate. Later they turn yellow and finally quite dark in color. There is seldom any apparent ropiness in the dead tissue as in the case of the other form of foul brood. Seldom is there a noticeable odor such as is so apparent in advanced stages of the American type of the disease. Queen and drone larvæ are usually attacked early. This is one of the common tests in early stages for determining which disease be present. The disease is usually more destructive in spring and early summer.

218 DISEASES AND ENEMIES OF BEES

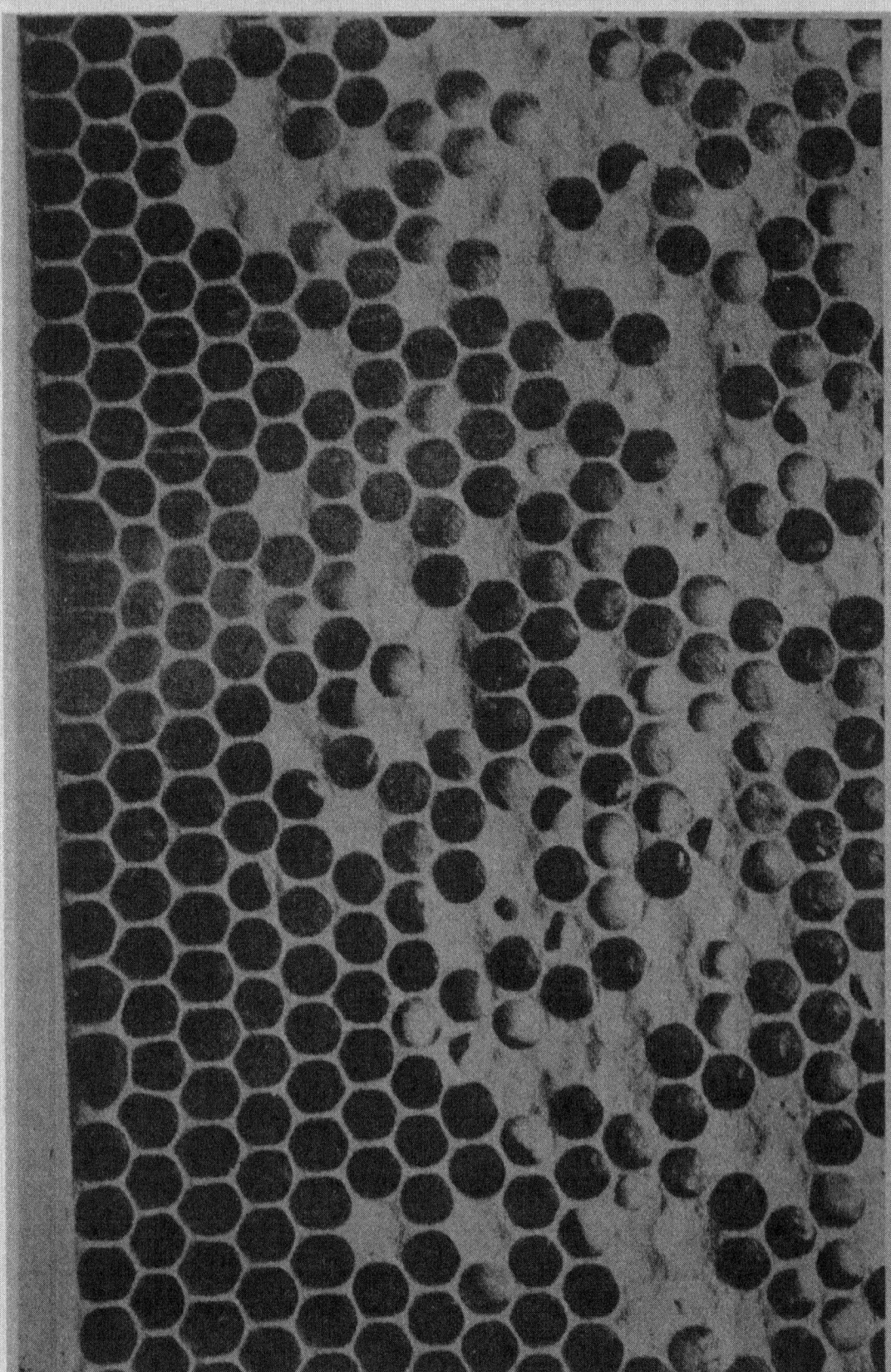

Fig. 105.—Appearance of larvæ affected by European foul brood.

Detection by Odor.—In some localities European foul brood is said to be attended with a decided odor, although unlike that of American foul brood, being more like that of decayed fish, according to Morley Petit, of Ontario.

Resistant Bees.—For some reason Italian bees seem to be much more resistant of this disease than the hybrids or blacks, and the best insurance against this malady is to re-queen all colonies with vigorous Italians. Some strains seem much more immune than others, so that it is desirable to secure a strain that has demonstrated its disease resistance.

Treatment.—There is much confusion on the part of the inexperienced bee-keeper between the two diseases, and since the treatment advised for one is entirely unsuited for the other it is important to make sure which disease is present. What is known as the Alexander plan is now generally regarded as the only dependable treatment for European foulbrood. The essential element of this plan is the saving of the combs instead of melting them up as is the case of the other disease. There are many modifications of the plan as proposed by the originator.

Alexander's Plan.—To begin with, the queen is removed from the diseased colony in order to check brood rearing. The bees being relieved of nursing young brood, turn their attention to cleaning out infected matter from the hive with the result that given a new queen a few days later they often remain free from the disease. Mr. Alexander believed it to be necessary for twenty-one days to elapse from the time the old queen was removed before the new queen began to lay. Of late many bee-keepers have found that under favorable conditions a much shorter time is sufficient. In order to be successful with this method several things must be borne in mind. First the new queen must be a vigorous young Italian. Then the colony must be very strong and the treatment must be given in early stages of the disease. If the combs are fairly rotten with decaying larvæ it is too much to expect that the bees will clean them up again. Hybrids or black bees are seldom, if ever, able to rid themselves of the disease in this manner. Dr. C. C. Miller, one of the best known authorities, is quoted as follows:

> I know there are those for whom I have great respect who have bitterly denounced the practice of trying to save the combs in treating European foul brood. In my first dealing with the disease I melted hundreds of brood combs. If I am forgiven I will never do it again. Please be sure to note that I am talking about European, not American, foul brood. The loss of the combs is not all there is of it. Indeed, I think that is the smallest part. The greater loss is from the set-back in the work of brood rearing. It seems to knock things endwise for weeks, if not for the season. Far less is the interference when egg laying is suspended for eight or ten days.
>
> I think I hear someone say, "But your treatment does not seem effective for you keep on having the disease, while with the orthodox method and the combs destroyed there's the end of it." Pardon me, that may be true with regard to American but not with European foul brood. I treated the disease after the most orthodox fashion, destroying, as I have said, hundreds of combs, and so far as I could see, the disease was just as willing to return as with the drastic treatment. I think I'd rather keep brood and combs.

With reference to the Alexander plan of treating this disease as practised by Dr. Miller and others, it is well to repeat what has already been said, that no method has ever been found for eliminating American foul brood without destroying the combs. This method applies only to European foul brood and *sacbrood,* a mild disease described later on.

European foul brood was long known as black brood and first appeared in the East. It was known in New York for a number of years before it made an appearance in the Central West. While there are still many localities where it is not present it has spread into practically all parts of the country.

Sources of Infection.—The spread of American foul brood in the neighborhood of the diseased colonies is usually by means of robber bees which visit those which, because of their weakened condition, are no longer able to defend their stores, and disease is thus rapidly spread. Every bee-keeper should guard against the robbing of weak colonies. In case a colony dies from disease the hive should be at once removed, the contents destroyed, and the hive and fixtures thoroughly disinfected.

Another source of the disease is the use of *second-hand honey containers.* A large part of the western honey goes to market in sixty-pound cans. These cans when empty are sold at a very low price and many bee-keepers are tempted to make use of them. Honey placed in these containers is sometimes later fed to the bees, or while being refilled they have often been visited by the bees, with the result that foul brood has been carried to the apiary, often in a minute drop of honey.

Bee-keepers have sometimes brought the disease home by *the use of purchased honey* for feeding in time of short supply of stores. The writer has been surprised at the extent of the complaint of the spread of disease from these two causes. There is so little to be saved by the use of a second-hand container that bee-keepers can hardly afford to take the risk. In case it becomes necessary to feed the bees, good sugar syrup should always be used unless the honey is known to be from apiaries that are free from disease.

The use of hives, frames, etc., in which bees have died is not safe unless they have been disinfected. While disease sometimes appears from some unaccountable source, the bee-keeper should take every possible precaution to avoid its spread.

The presence of foul brood in an apiary is a serious matter to the owner and cannot but result in serious loss. Frequent reports come to the writer of the loss of entire apiaries, sometimes of many colonies, from foul brood.

MINOR TROUBLES

Sacbrood has long been known by the name of pickled brood. The name sacbrood is, however, much more appropriate because the dead larvæ do not melt down as they do in foul brood, but rather retain the full size, the body wall retaining the contents in the form of a sac. This disease is seldom serious in nature although it is mildly infectious and may be spread from one colony to another. As a rule no attention need be paid to it, as it usually disappears with the coming of a good honey flow.

If the queen at the head of the diseased colony be old or failing it is well to re-queen with vigorous young stock. Some recommend treatment for this disease as with foul brood, but that is seldom, if ever, necessary. In bad cases where the colony is weak the queen should be replaced and the colony strengthened by the addition of frames of emerging brood.

Symptoms of Sacbrood.—This disease somewhat resembles European foul brood and is frequently confused with that disease. Inspectors have in several instances been called long distances to deal with European foul brood, only to find after a few days' delay that the disease was sacbrood and had nearly disappeared of itself. The dead larvæ are found to be extended the full length in the cell with the sharp pointed end slightly turned upward. The dead tissue soon assumes a dark color and there is little or no odor to the combs.

Chilled or Starved Brood.—The young bees die from several other causes than any of the three diseases mentioned. It frequently happens in early spring that the brood nest expands rapidly during the first warm days, with the result that a sudden drop in the temperature makes it impossible for the bees to warm all the brood and a portion becomes chilled. The supply of honey or pollen is also at times exhausted when unfavorable weather conditions make it impossible for the bees to gather stores for a considerable period and much brood is lost from starvation.

When bees are being moved without sufficient ventilation the brood is sometimes lost from overheating. All of these causes are occasionally responsible for the supposition that foul brood is present when, in fact, it is not.

Poisoning.—The growth of the commercial fruit industry has developed a new difficulty,—poisoning the bees. It frequently happens that some overzealous fruit grower, blind to his own best interest, sprays his fruit trees while in full bloom. This not only injures the chances of getting a full crop of fruit, by washing off the pollen at a critical period, but results in the destruction of the honey-bees whose presence just at this time is very essential to his success. So serious had this condition become in some localities that laws have been passed to prevent the spraying of fruit trees during the period of full bloom. It would seem that any man who is sufficiently progressive to spray his trees would realize the fact that he can get better results in spraying immediately after the petals fall.

Dysentery is usually caused by too long confinement or poor stores. Under normal conditions the worker-bee voids her excrement only when on the wing. When long periods of time elapse during which they are unable to fly and thus relieve themselves of the accumulated waste in the intestines, they are sometimes compelled to discharge within the hive. When this condition is reached they soon die, unless a change in the weather permits them to get out and to clean up. Under such circumstances the combs are badly soiled and the bees die amidst the filth.

Honey-dew or other poor stores is quite likely to cause this trouble. It is an important matter to see that the hives are supplied with honey of the best quality for wintering, especially in the North where the bees are confined for weeks or months without flight. (See Chapter XIII.)

Weak colonies are more susceptible to dysentery than strong colonies, for the reason that a greater amount of food will necessarily be consumed in order to keep up the heat, and the waste is consequently greater.

Prevention.—From the above it will be seen that dysentery is generally a winter disorder and that proper wintering insures freedom from the trouble. While strong colonies, with good stores and proper protection, seldom are seriously troubled, still even they may have trouble under unfavorable conditions, or during very long confinement.

Remedy.—About the only remedy is a good flight on a warm day. If the bees are beginning to show signs of this trouble in the cellar in winter, and a warm day comes which will permit a safe flight, it will pay the bee-keeper to set them out and to put them back again at night after they have returned to the hive.

Mice.—The mice sometimes enter the hives in winter, either in cellars or out-door wintered colonies. The author once saw a hive where the little rodents had gnawed through an old bottom board and really had destroyed the colony by eating the combs and disturbing the bees during their winter rest. Both the white-footed wood mice and the common house mice are likely to cause such mischief. Mice and rats are also very destructive in the honey house by destroying surplus combs, sections, etc., and it is well, if possible, to make the honey house mouse proof.

Skunks.—The normal and preferred food of the skunk is insects and mice. It is then to be expected that bees will suffer where skunks are common. They sometimes learn to scratch at the entrance of the hive and to catch the bees as they rush out. Skunks are also fond of honey, as the writer has found by feeding it to these animals in confinement. However, they are unable to get at the honey in the hive and the only injury from these animals is to the bees.

Skunks are of considerable value in keeping down the number of rodents and such insect pests as grasshoppers and crickets, and where they are not too abundant should be encouraged. It is frequently wiser to protect the bees and poultry from the animals and leave them free to war on rats and mice than to destroy the skunks and have to fight the other pests. It is less trouble to guard against the skunks. In localities where they become over-abundant, it may sometimes be advisable to destroy them (Fig. 106).

Dragon flies, mosquito hawks, snake feeders, or devil's darning needles, all of which are common names for the same insects, are sometimes mentioned as enemies of bees. The trouble seems to be limited to restricted localities, and while there are sometimes instances where individual bee-keepers suffer considerable annoyance from these insects, especially from the loss of young queens

THE WAX MOTH

which are caught on their mating flights, the insects cannot be said to be generally injurius.

The robber fly is a large insect that flies with a loud buzz. It is a rapacious fellow, seeking those it may devour. Butterflies, bees, grasshoppers, and even wasps and beetles fall victims to its voracious appetite. It is seldom sufficiently abundant to cause appreciable injury in the bee yard and may be regarded, on the whole, as a useful insect (Fig. 107).

Fig. 106.—The natural and preferred food of the skunk is insects. The honey-bee is a tempting delicacy to the skunk palate.

Spiders also sometimes weave their webs in situations where the heavily laden bees fall into them and are lost. Large webs in the immediate vicinity of the hive should not be tolerated, but aside from that little is to be feared from spiders.

THE WAX MOTH

The larger wax moth (*Galleria melonella*) is very widely distributed and among indifferent bee-keepers is a serious pest. It is present in nearly all portions of Europe and North America where bees are kept, excepting the high altitudes of Colorado and other western States. In the vicinity of Denver it has several times been introduced, only to disappear within a short time, apparently being unable to live in the high and dry atmosphere of that region. It is more destructive in the warmer parts of the country than in the northern sections where the season is not so long.

The adult is an inconspicuous little moth of grayish color, quick to take flight on the opening of the hive (Fig. 102). They remain secluded during the day unless disturbed, but are apparently very active after nightfall. The eggs are laid in crevices in or about the hive where the larvæ can readily find their way to the combs. The insect is very prolific and once a weak colony becomes infested the total destruction of the combs may be completed in but a short period of time.

FIG. 107.—The robber fly. (After Washburn.)

Concerning the laying, Paddock[1] says:

In the cages where empty comb was supplied, the eggs were always laid in cavities and if possible in such cavities as were well protected. Only one egg is deposited at a time, though in working over a small piece of comb the eggs may be placed close together, apparently in masses. The eggs are always securely glued to their resting place; usually the shell will break before the egg is loosened. The number of eggs deposited by one female has not been ascertained but moths which had not deposited eggs were killed and the eggs in their ovaries counted. The average number of eggs counted was 1014. The time consumed in laying the full quota of eggs varies with the generation, averaging nine days in the first and seven days in the second.

When first hatched the larvæ are white and very small. They burrow into the combs at once. The larval period is spent in

---

[1] Observations on the Bee Moth, Journal of Economic Entomology, vol. vii, No. 2.

tunnelling through the combs along the midrib. The pollen stored in the cells, as well as the wax of which the combs are constructed, seems to furnish them with food. The tunnels are lined with silk similar to that of which the cocoons are composed. It is not long until the combs are but a mass of webs and refuse (Fig. 103). The length of this stage varies from thirty-five days in the second brood to forty-five days with the first brood according to the author above quoted.

The cocoons are spun in masses under the cover, behind the ends of the frames or in any other situation seeming to offer protection, but usually within the hive where the larval period has been spent. About two weeks are required to complete this stage, after which the adult moths will appear.

In the extreme South it is probable that breeding continues throughout the year with little interruption. In the North only such individuals as are fortunate enough to select a place free from extreme cold will survive. Those remaining in hives in the open air in which the bees have perished will likewise die before spring, as they cannot endure severe freezing. There are always a few tucked away snugly in the hives near the clusters of bees, which are kept sufficiently warm to insure their safety. These will shortly populate a large area with their offspring when warm weather comes. They are also able to pass the winter in empty hives that are carried into the cellar or other place where the temperature does not drop much below freezing.

The Remedy.—The wax moth may be regarded as a symptom that something is wrong, for a normal colony of bees will usually defend themselves against this pest without difficulty. Italians, however, are better able to contend with it than the common strains. The blacks are especially liable to succumb to an attack of moths. Usually it is the weak and queenless colonies that fall victims to its ravages.

Three adult moths may be seen on the comb in Fig. 102. The larvæ are repulsive caterpillars and reach an inch or more in length. Fig. 103 shows the work of these insects in a little more advanced stage. If left undisturbed such a comb would very shortly be entirely destroyed.

Amateur bee-keepers frequently complain that the moth is destroying their bees and inquire what to do for it. The answer is: Keep your colonies strong and replace old and failing queens with vigorous young Italians. Colonies that have become weakened by disease fall an easy prey to the moths.

Experienced bee-keepers are seldom heard to complain of this trouble, for they have long ago learned that constant vigilance is the price of success, in the apiary as elsewhere. The bee-keeper who does not examine the brood nest of his colonies occasionally has no means of knowing the condition of his bees. Frequent examination will enable him to detect and avoid the conditions that provide favorable surroundings for the moths.

Care of Empty Combs.—The moths are a source of annoyance to the bee-keeper who has large numbers of empty combs during a part of the year. During the warm months there is no better place to store empty combs than over a strong colony of Italian bees. After the season is over and cold nights come they may be placed in any cold place safely, for freezing will effectively check the work of these insects. It is well to have a tight compartment where no adult moths can get in so that they will be safe after warm weather comes again. There is always danger in putting away empty combs in warm weather, that eggs may be present and that the combs may be destroyed before the presence of the insects is discovered.

When combs either empty or containing honey are found to be infested with moths they should be cared for without delay as the insects develop very rapidly.

If only a few combs be injured they may be placed in strong colonies and the bees will clean them up quickly and effectively, throwing the dead larvæ at the entrance of the hive within a few hours. If there is a large number of combs it is well to place them in a tight room and fumigate them. This may be done by putting a quantity of sulfur in a dish, first pouring alcohol over it so that it will burn readily, and setting it on fire. Care should be taken to place the receptacle containing the burning sulphur in a safe place on a large stone or metal, or in a larger tub or pan containing water. The building should be closed very tight to prevent the escape of the fumes. The combs should be separated to insure the fumes reaching all parts. Sometimes a second or even a third fumigation will be necessary to insure the destruction of all of the caterpillars in bad cases.

Bisulfide of carbon may be used to accomplish a similar result, but great care is necessary as it is highly explosive and dangerous. In the use of this drug the combs are placed in a tight closet or box and a quantity of the liquid placed in an open dish above them. It evaporates rapidly and the heavy fumes settle over the combs, thus effectively killing the moths. No fire or light should be allowed about when this liquid is being used.

LESSER WAX MOTH

There is a very small moth whose larva sometimes becomes troublesome in comb honey. It is not nearly so destructive as the larger species and its work is seldom noticed excepting in the comb honey. It frequently appears in honey that has been a considerable time in the market and greatly injures the appearance by spreading its webs over the cappings and making its small burrows into the wax, thus causing leakage, waste, and a bad appearance. The remedy is fumigation.

THE BUSINESS OF AN INSPECTOR

More than half of the States and several Canadian provinces now have inspectors with police powers for the purpose of controlling bee diseases. But a few years will elapse until every State and province where bee-keeping is an important industry will make such provision. Where the work is thoroughly done a number of men are required to cover the field, so that the inspection work is growing in importance and in opportunity.

With the appearance of bee disease it was very natural for the bee-keepers to look for assistance from the State. Alone the bee-keeper is helpless against infection from uncared for apiaries. He may be ever so careful and efficient, but without protection from unnecessary contagion he must carry on the fight against disease for a long period of time, move his apiary, or go out of business. Since bee-keeping is being developed as a specialty on which many have come to depend for a livelihood, it is imperative that legal protection be extended.

The sole thought in the beginning was to provide for the examination of all bees and to compel proper treatment or destruction of those found to be diseased. The inspector was given no choice but to examine all the bees in the localities to which he was called. At the same time funds sufficient to examine but a small part of the bees in any State were provided.

Of late the tendency has been to depend more and more upon proper instruction. Until much larger appropriations are available it will not be possible to reach a large percentage of the bees in any State. If the bee-keeper is an intelligent man, an hour or two of the inspector's time is all that he will require. If upon examination one or more colonies are found to be diseased, the inspector will be able to point out the characteristics of the particular disorder and to give proper instructions for its treatment. The bee-keeper will then be able to recognize the trouble when he finds it in other colonies and to deal with it promptly. It would hardly seem to be the province of the State to examine every colony and give the necessary treatment. If such a plan is followed a week will often be necessary to deal with a single large apiary.

Where the owner is careless or indifferent it will become necessary for the inspector to be very thorough in his examination and to insist on proper attention to diseased colonies. Police power is necessary because of the fact that many persons who keep bees are so ignorant of their care in either health or disease that they cannot be convinced of the necessity or value of proper attention. In such cases the bees will be left to menace the surrounding apiaries until such time as they shall finally succumb to the disease.

Requirements for Successful Inspectors.—No man should be intrusted with police powers who does not have proper regard for the rights and feelings of those with whom he is required to deal. He should be able to meet a trying situation and to reason with those who are disposed to resent his visit. Fortunately most bee-keepers are coming to be very anxious to learn of the presence of disease on its first appearance in their apiaries and will communicate with the inspector at the first suspicious sign. In such cases the inspector will be welcomed and information will be gratefully received. However, when disease is found it becomes necessary to examine other nearby apiaries to ascertain to what extent the disease has been spread. Many of the bees will be found in boxes, kegs, or hives where the combs are built crosswise for lack of foundation. The conditions are such as try the patience of a mild-tempered man, and to ascertain the condition of the colony and leave the owner in good temper requires the exercise of much skill and diplomacy.

If the inspector is able to give the owner of such bees encouragement and advice about proper care of bees without offence, his visit has been of value aside from the possible check of the spread of disease. The time bids fair to come very shortly when the inspector's field shall be broadened until his duty will be to instruct in the general management of the apiary as much as to find disease. The great difficulty with present laws lies in the fact that no man who is not a well-informed beekeeper is competent to deal with disease. The inspector's instructions regarding disease will be imperfectly understood by the box hive bee-keeper, nine times out of ten, and if he undertakes to treat his colonies himself he will destroy them or scatter the disease instead of checking it. It thus becomes necessary for the inspector to personally supervise the treatment or destroy the diseased colony. A diseased colony in anything but a modern equipped hive is worthless, as it will cost more to transfer the bees, as a rule, than a diseased colony is worth.

The man who is fully informed concerning up-to-date methods of bee-keeping will be able to handle disease in his own apiary if he can be protected from further infection. The problem then becomes one of making every man who keeps bees an up-to-date bee-man. In localities where disease gets well established it will be impossible to eradicate it entirely until every bee-keeper becomes expert. Disease has the effect of making expert bee-keepers anyway, for those who do not become proficient are likely to lose all their bees within a short time.

The bee inspector is usually regarded as the official representative of the industry and should be able to represent it creditably under any circumstances. It is not enough to be informed concerning detection and treatment of disease, but he must be able to deal with problems relating to any branch of bee-keeping. Bee-keepers whom he visits will give him their hardest problems to solve and people in other walks of life will turn to him with any question relating to the business. He will be called upon to give expert testimony in case of litigation involving bee-keepers and to settle various disputes between persons where the rights of one or the other are in question.

Opportunity.—The various State agricultural colleges are rapidly taking up bee culture, and it bids fair to take its legitimate place in the college curriculum. Within a few years the inspection work, instead of being under direction of a separate State department, as now in many States, will be organized in connection with extension work in bee-keeping. As the business of bee-keeping is taking on new life the demand for properly equipped men will probably exceed the supply for several years to come. That this condition has not developed sooner is because the bee-keepers have been slow to recognize the great advantage that would come to the industry as a result and to demand the same recognition given other lines of agricultural activity. A few who have not caught the spirit of the times are loud in their complaints that for the agricultural colleges to take up bee-keeping will make too many bee-keepers and that there will be no market for the product of the hives. Fortunately they are now in the minority and progressive bee-men are in the lead.

It would seem that all that is necessary to meet such complaints is to point to the increased profit that has come to the dairy and other farm industries by such development. With better methods and larger production has come better markets and higher prices. The same condition will apply to bee-keeping, which is just beginning to come into its own.

QUESTIONS

1. What conditions indicate the presence of American foul brood?
2. Describe the method of treatment.
3. How does European foul brood differ from the American type?
4. Discuss methods of treating this disease.
5. How is foul brood spread?
6. What is sacbrood?
7. Are dragon flies and robber flies serious enemies of the honey-bee?
8. What other enemies must the bee-keeper combat?
9. Discuss the cause and prevention of dysentery.
10. How can the wax moth be controlled?
11. Describe the necessary care of empty combs.
12. Why is a bee inspector a necessary officer?
13. Discuss his duties and opportunities.

CHAPTER XIII

WINTERING

With the rank and file of bee-keepers in the Northern States, the wintering problem is the most serious one they have to face. In some localities brood diseases may be a serious menace for a time and cause great losses, but the wintering problem must be met in all sections of the North and must be faced every winter. While many professional bee-keepers have learned to prepare their bees for winter so carefully as to meet with little loss, the average small bee-keeper suffers seriously from this cause and in severe winters occasionally loses a large part of his stock.

In the Southern States the problem is a somewhat different one. In some parts of the South instead of being a question of suitable protection from cold, it becomes a question of checking brood rearing during the period when no honey is to be gathered and providing sufficient stores to bring the colony to the next honey flow in good condition. When stores are short the colony will delay brood rearing beyond the time when large numbers of young bees should be hatching in the hive, with the result that the first period of profitable honey flow is passed before the colony becomes strong enough to make the most of the opportunity.

In the high altitudes of Colorado and the West it is a common practice to winter the colonies in the open air without extra protection. While in these high altitudes with the prevalence of sunny weather the bees can fly so frequently as to insure a large portion of the colonies coming through the winter alive, it would seem that there must be an unnecessarily heavy mortality among the bees and that with suitable protection there might be considerable saving in both bees and stores.

Essentials of Successful Wintering.—It is common to speak of wintering as though proper protection from cold were all of the problem. In fact at least two other things are of more importance: first, of course, a supply of suitable food large enough to last until the flowers bloom again; next a vigorous young queen. After these, suitable protection should be considered.

When the bees are unable to fly for long periods of time, as in winter, proper food is of great importance. Normally the bee voids its excrement only while on the wing. The wastes that accumulate in its body during the long weeks of inactivity are a severe tax at best and with low-grade food stores, the quantity becomes so great as to swell the abdomen to the point of causing death. In mild winters when there is frequent opportunity for cleansing flight, bees wintered out of doors will go through safely on almost any kind of stores if the quantity is sufficient. There will be, however, a much greater mortality among the bees on poor stores than on those of good quality. Bees wintered in cellars, or outside in severe winters, cannot be expected to come through in good condition on poor stores, even though they survive at all.

White Honey the Best Winter Feed.—The whiter the honey, as a rule, the less waste it contains and there is no better winter feed than white clover honey. The color is not always a safe guide, however, for some aster honey is said to be light in color and aster honey seldom gives good results as a winter feed.

The dark fall honey, especially when mixed with pollen, is much less desirable, and honey-dew is disastrous. It is a common plan among practical apiarists to extract all late honey, which has not had time to be thoroughly ripened, from the combs at the close of the honey flow and to replace with sealed white clover honey, or to feed sugar syrup. The best grade of granulated sugar should always be used for this purpose as it makes a very good substitute for honey for wintering. Equal parts of sugar and water are frequently used, although best authorities recommend less water; three parts sugar to two parts water, or two parts sugar to one of water being regarded as better. When it becomes necessary to feed from lack of sufficient stores or to replace unsuitable stores, it should be attended to immediately after the close of the honey flow to give the bees time to get things in readiness for winter before the first cold snap. (See Chapter VIII, Feeds and Feeding.)

Failing Queens and Old Bees.—The old bees that have gathered the year's honey crop will all die before the opening of the next season's harvest. It is very important, therefore, that the coming of winter shall find large numbers of newly hatched bees to replace them. It is the late hatched bees that are not exhausted by honey gathering that survive the winter and begin the work of the following season. The bee-keeper should see to it that conditions favor brood rearing in the fall to insure this condition.

It often happens that a colony which has been strong all summer and perhaps has stored a large surplus will die during the winter or early spring from the failure of the old queen. It is important that the bee-keeper see that all colonies have vigorous queens at the time of preparing for winter. All colonies that cast swarms during the season will have young queens, if they have any at all, as the old queen always leaves the hive to go with the swarm. For this reason it often happens that one will get a new swarm only to find it dead or worthless the following spring. The bees usually replace a failing queen, but they cannot always be depended upon to do so. When the queen begins to fail in late fall or winter, conditions are not favorable for rearing another and if a virgin is raised at this season she has no opportunity for mating, so is worthless.

Influence of the Queen.—It should be understood that the queen herself does not have a direct influence on the wintering of the colony. In fact she might be removed entirely and if other conditions are right the colony will come through safely. The importance of having a vigorous young queen lies in insuring plenty of young bees at the beginning of winter and that brood-rearing will commence in due season in spring. Colonies with failing queens are likely to be so badly weakened before their true condition is discovered in spring that they will be worthless or nearly so.

Practical bee-keepers look very carefully after the queens in making winter preparation. Some apiarists re-queen all colonies every year to insure only young queens. This method results in the destruction of many valuable queens, however. It is a common practice to re-queen every other year, thus keeping the queens for two years, while others keep a record of every colony and only replace the queens when they show signs of failing. If bees are on straight combs in movable frame hives, as they must be for profitable care, it is an easy matter to remove the old queen. She must always be removed before a new queen is given. Otherwise the bees will destroy the newcomer. Queens are for sale by numerous queen breeders who will supply them from April to October. Directions for introducing them come with the little cage in which they are mailed. This subject is further considered in Chapter VII.

Protection from Wind.—Not all of wintering lies in getting the colonies through the winter. It is equally important that they come through in such condition as to build up early, in order that every colony be very populous at the beginning of the honey flow. The changeable weather of early spring must be considered and some protection be provided against the chilling winds of this season. As soon as warm days come, the queens will begin to lay in earnest. Within three days from the time the eggs are laid the larvæ hatch and require a very warm and even temperature. Baby bees are even more sensitive to unfavorable conditions than baby chicks. It often happens that a few warm days will result in the appearance of considerable quantities of brood in the hive. A sudden drop in the temperature makes it difficult for the bees to keep the brood nest sufficiently warm, with the result that a part of the brood is likely to be chilled and consequently lost. Every possible means should be used to save the energy of the colony at this season.

Too much value can hardly be placed upon a good windbreak. Evergreens so planted as to break the wind from the north and west are very good. The author's apiary is sheltered by a blackberry thicket immediately behind the hives and back

Fig. 108.—The value of a good natural windbreak behind an apiary can hardly be overestimated.

of that is a grove of native trees (Fig. 108). The apiary was formerly in the grove where the wind swept under the trees. The difference in the condition of the colonies in spring, since moving to the new location, is surprising indeed.

When brood rearing commences the bees require quantities of water and this accounts for their frequenting the watering troughs so freely in early spring. Water should be placed near at hand to save long flights in search of it. A tub, trough, or other receptacle partly filled with shavings, chips or the like to enable the bees to get the water without danger of drowning should be provided. (See Chapter IV.)

Protection in Spring.—Many bee-keepers complain that after they bring their bees through the winter in the cellar they lose a large part of them through the spring, the stock dwindling after being placed on the summer stands. Several things might be the cause of this condition. Too many old bees, or colonies that went into the cellar weak, or lack of suitable protection might be responsible. It is important not only to place the bees in a carefully sheltered position after they are removed from the cellar, but in addition to provide some protection in the way of packing.

Over large areas the principal flow is from white clover, which is of comparatively short duration. It is only the colonies that are strong in bees at the beginning of the flow that will return substantial profits to their owners.

Strong Colonies Also Essential.—To the above general principles we must add another—strong colonies. While it is sometimes possible to winter a weak colony or even a nucleus, it is seldom worth while. If a colony is weak at the beginning of winter by the time spring arrives there is not likely to be enough bees left to build up without the addition of brood or bees from a stronger colony. It would be wiser to unite several weak colonies to make one vigorous one than to bother with the weaklings.

A strong colony of bees will require less honey to winter successfully than a small one. The source of heat is the food consumed and the larger the cluster the more animal heat will be retained.

METHODS OF WINTERING

In considering the various methods of wintering here presented the reader will bear in mind that some methods suited to the latitude of St. Louis would not be safe for northern latitudes. Chaff hives, paper cases, and similar methods which are entirely satisfactory for Southern Missouri and southward are not to be recommended for Minnesota, Wisconsin, Ontario and similar sections. Outdoor wintering, however, may be safely practised as far north as Canada if proper precautions are taken in the winter preparations. The success or failure of outdoor wintering in any latitude will depend to some extent on the surrounding conditions, such as windbreaks, as well as the actual protection of the hives. What has been said about the desirability of spring protection for cellar wintered colonies will apply with equal force to colonies wintered outside.

Paper Cases.—One of the common plans for outdoor wintering in the southern part of the region where winter protection is necessary and one that brings fairly satisfactory results in ordinary winters is the paper case. Tar paper or other black paper should never be used because of its tendency to absorb heat. The hive under a black protection case will suffer from such extremes of heat and cold as to render it worse off than though it had remained without protection. Light colored building paper, however, will answer very well.

To make such a case two or three corn cobs are laid over the top of the frames and a cotton cloth or burlap spread over them. The purpose of the cobs is to permit the bees to move freely from place to place to reach their stores. An empty super is then placed on the hive and filled with dry leaves or chaff. The cover is then placed on the super and the whole covered with several layers of newspapers. A large sheet of heavy building paper or other waterproof paper is then placed over all and folded around the hive and fastened as shown in Fig. 109. The hives shown in the illustration are without the super of leaves. There is a disagreement among bee-keepers as to the value of this porous packing material over the frames. The purpose is to absorb the moisture during cold weather. Some argue that the bees are safer without it, but the author is a believer in absorbent cushions. Building paper cases are at best a scant protection.

The winter of 1911–12 caused such heavy losses among out-of-door wintered colonies as to discourage many advocates of outside wintering. This was an extraordinary winter with unusually low temperatures and long periods between days warm enough for a flight. Investigation shows that a large per cent of the loss in this unusual season was due to poor stores and careless preparation for winter. The two preceding winters had been so favorable that many bee-keepers were inclined to

Fig. 109.—Paper winter cases are at best scant protection, but are good for cellar-wintered bees after they are placed on the summer stands.

take the risk rather than go to the trouble of careful preparation. The losses have not been without compensation, for the result will not be soon forgotten, and the bees will receive better attention at the proper time for years to come.

Outdoor wintering is very successful in the hands of some of the most successful and extensive honey producers. In fact a few have practised no other method for many years and get uniformly good results. With outdoor wintering it is very essential that great care be used to see that stores are of sufficient quantity and of good quality.

One of the principal arguments in favor of cellar wintering is the saving in stores, which is considerable. In general it is estimated that from one-third to one-half more honey will be consumed when wintered outside. This is offset by earlier brood

Fig. 110.—The Dadant method of outdoor wintering in large hives is suited to localities where the bees have frequent flights during the cold months.

rearing and a generally stronger condition when properly wintered out-of-doors.

The Dadant Method.—The Dadants are extensively engaged in honey production in Hamilton, Illinois, directly across the river from Keokuk. They use a large hive and cover the brood frames with a straw mat. Woven wire is tacked to one corner of the front of the hive and then made to encircle it on both

PITTING OR BURYING

sides and the back. The space thus provided is filled with leaves. The front is provided with no protection. Fig. 110 shows the method employed by C. P. Dadant, one of the best known American bee-keepers, of wintering in the Dadant hive which he has used for many years. The results seem to be satisfactory when proper stores are supplied. This way, while suited to the conditions of Keokuk and southward, would hardly be safe much

FIG. 111.—One method of packing on the summer stands: (a) roof of composition materia (b) board over entrance.

farther north, judging from the reports of outdoor wintering from northern sections. This plan is not suited to small hives, such as the eight-frame Langstroth so commonly used.

Pitting or Burying.—On sandy or other very porous soils a few bee-keepers practise pitting or burying. A trench is dug about eighteen inches deep and 2 × 4's placed in the bottom to keep the hives off the ground. After its bottom has been removed the hive is placed on the scantling and the cover slightly raised to provide for upward ventilation. About eighteen inches of straw is placed over the hives and this in turn is covered with a layer of loose earth. Not over two layers of hives should be placed in such a trench. There are only a comparatively few locations where this method is suited to the conditions, as a well-drained situation and porous soil are essential. Bees pitted in this manner are liable to be disturbed by skunks or other burrowing animals and serious loss to result. There is danger of loss also if they be buried either too deep or not deep enough. While the method may do as a makeshift under temporary conditions it is not to be generally recommended.

Fig. 112.—Parts of a double-walled hive.

Packing on Summer Stands.— Another method suitable for southerly latitudes where only slight winter protection is needed is packing on the summer stands. The illustration (Fig. 111) shows an apiary in southern Iowa. In this case a tight board fence about thirty inches high is used as a windbreak to the north of the bees. The hives set in a long row about six inches from this fence. Leaves are packed between and behind the hives and a waterproof roofing is placed over all to shed the rain. In this plan the same method of packing empty supers with leaves over the brood nest is used as described under paper cases. Various modifications of this plan are in use throughout this latitude from Kansas and Missouri eastward.

Chaff or Double Walled Hives.—Too many bees are left without attention in single walled hives and for the average small apiarist double walled hives similar to that shown in **Fig.**

Fig. 113.—Double-walled hive assembled. The double-walled hive with the space between filled with chaff or cork is suited to the conditions of localities where winters are mild.

112 would be far better. The space between the two walls is packed with chaff and over the brood nest is placed a tray also filled with chaff and a large cover telescopes over all. Fig. 113 shows the hive as it appears in use. During winter the entrance is contracted to a four inch width. There are several hives on the market built on this plan.

The chaff hive like the foregoing methods is not well adapted to the far northern regions and much of the complaint against this method of wintering is perhaps from regions where it should not be used. For intermediate latitudes, with good stores and proper attention, this hive should be expected to give satisfactory results. Extensive bee-keepers who have used a similar hive in Michigan for many years report that the average loss has not exceeded ten per cent. The practical bee-keeper should not be content to follow any plan by which he could not reduce the winter losses below this figure. A ten per cent loss is sufficient to condemn any system. However, a large part of winter losses by any method is always to be charged to failing queens and improper stores, two things which can hardly be charged to the method of protection.

There has been much discussion as to whether with the use of double walled hives there should be a sealed wood cover under the tray of chaff. The best authorities take directly opposite views on this subject, some holding that there should be no upward ventilation and the other side holding that upward ventilation is essential to absorb the surplus moisture in extremely cold weather. The author, as already stated, holds to the latter view, and in practice uses it as well as recommends it.

The double walled hive is a most excellent hive for early spring use, as the bees are not subject to such sudden changes of temperature as is the case in single walled hives.

The large size and extra weight are against the double walled hive for use in extensive apiaries where every part should be interchangeable. For the use of comb honey producers, however, who have less occasion to manipulate the hive bodies, there is not so much objection to be made. There can be no question but that there will be a greater saving in stores for early spring brood rearing in a double walled hive than in the ordinary single walled one. More honey will also be stored from fruit bloom and other early sources as a smaller number of bees will be required to maintain the required heat for brood rearing in the hive. Taken altogether, there are many advantages from its use to offset the greater weight and bulk.

One of the greatest advantages is in leaving the hives in one position the entire year. Winter preparations require but a few moment's time with each hive. All that is necessary is to contract the entrance, put the inner cover in place, place the tray of chaff in position over the frames, and place the telescope cover over all and the job is done. The busy man who has but a few bees for diversion and who wishes to be relieved of unnecessary manipulations in caring for them will find the double walled hive to be ideal for his use. In fact the author feels that it is the best possible hive for amateurs generally, who do not keep more than twenty-five to fifty colonies. As to whether it will pay the large producer to use this type of hive is not quite so evident. Some find them satisfactory on an extensive scale, while others feel that they are not suited to the use of the extensive honey producer.

Packing Cases.—Various kinds of packing cases have been in use for many years, so the idea is not new. However, the tendency of the time is to abandon cellar wintering in favor of packing cases. With proper preparation bees will be safer in winter cases than in a cellar and will reach the season of honey flow in better condition than by any other method of wintering.

Bees are successfully wintered in packing cases as far north as Canada, and some of the most extensive honey producers have abandoned expensive cellars for their use. The most common plan is to pack four colonies in one box with entrances facing two to the east and two to the west, or two to the south and one each to the east and west. Less labor is required to prepare the boxes with only two openings. North openings are not to be recommended.

Where four colonies are packed in a case, two sides of each hive have the additional protection of other hives warm with the clusters of bees. The colonies will thus be much warmer than when packed singly.

Several years of observation indicate to the author that bees winter better in larger hives than in smaller ones. Unless the hive is at least as large as the ten-frame Langstroth, packed colonies should be wintered in double stories, or with a deep bottom or empty super underneath. The Minnesota combination bottom and feeder is used to some extent for this purpose and is highly recommended by those who have tried it.

It is unfortunate that the eight-frame hive is in such com-

Fig. 114.—Packing box with hives inside ready for leaves or other packing material for outdoor wintering. (Received from Iowa Agriculture College.)

mon use. While the eight-frame hive is good in the hands of expert comb honey men, the larger hive is much better for ordinary use. For wintering in eight-frame hives considerable difficulty is sometimes encountered to get enough honey into a single hive-body to insure sufficient stores. Good results have been secured by placing two hive-bodies one above the other and leaving about ten or fifteen pounds more honey than seemed necessary. Two double-story hives are then placed side by side, close together, in a drygoods box of one inch lumber, or four in a packing case. The drygoods and clothing merchants get a number of boxes every fall just about the right size to pack two colonies together. A box can sometimes be secured large enough to pack three colonies side by side (Figs. 114, 115 and 116), but

Fig. 115.—Packing two colonies with dry leaves in a goods box. The entrances are left open to give the bees opportunity to fly on warm days.

these do not give as good satisfaction, for the bees from the hive in the middle seem to enter the hives on either side, until the colony which is most favored as far as warmth is concerned comes through the winter weak from loss of deserting bees.

The advantage of using the drygoods box lies in the lower cost and less labor necessary to get it ready for a packing case. Sufficient lumber to make such a case would in most localities cost several times as much as is paid for the boxes. They need so little alteration that but a few minutes is necessary to make one over. As will be seen by Fig. 116, a six-inch strip is removed in front of the entrance and turned inside the box to prevent the packing from dropping down in front. On warm days the bees are free to fly. Dry leaves are used for packing and about four to six inches of space is filled all round the hives and usually

Fig. 116.—Snug for the winter.

from ten to twelve inches over the top, the more the better. As the hives are two stories high the bees have an abundance of room for spring brood rearing.

As before stated there should be an abundance of honey. With a surplus available in the hive in spring and the hives protected from the cold winds by the packing, they need not be opened until the beginning of the honey flow. Colonies thus packed, and opened for the first time about the first of May, have been found to be full of brood and honey from fruit bloom.

Sometimes queen cells will be started very early in preparation for swarming. At the same time colonies without protection were making slow progress toward building up.

It is very apparent that such colonies as described on May first are worth much more as honey gatherers during the clover flow. When colonies reach this stage sufficiently early it is sometimes possible though seldom advisable to make increase ahead of the clover flow. Where the main flow is later in the season, this extra early brood rearing is not so important, though the colonies should be strong. Over large areas of the Northern States the bees need careful attention to build them up early enough, as a rule. In these packing cases the bees will sometimes store surplus from fruit bloom and dandelion.

Reports of success from wintering in these or similar cases are uniformly good where the work has been properly done, over nearly all the States and Canada.

For large apiaries a case which holds four colonies, two facing east and two west, is perhaps more desirable. The worst objection to packing cases is the large amount of labor in preparing for winter and the bother of storing the cases in summer. As one bee-keeper expressed it, the results were good but it required acres of space to store his packing cases in summer. Instead of nailing the cases into permanent form it is a common practice to fasten at the corners with hooks, so that the parts can be piled up compactly during the summer months. In large apiaries the use of drygoods boxes is hardly practical because of the difficulty of disposing of so much bulk when not in use. Where only a few are in use they can readily be turned to account as chicken coops in summer (Fig. 117).

The packing case is perhaps the safest and most generally successful of any method of wintering, taking the country as a whole. It is adapted to any section, north or south, and permits the bees to fly wherever the weather is sufficiently warm. There is much trouble from outdoor wintered colonies losing large numbers of bees which fly out on bright days when it is too cold for successful flight, dying upon the snow. In the packing cases this trouble is avoided as the bees do not feel the heat of the sun until the air is sufficiently warm to permit a safe return. Of course there will always be some old bees which will die outside after every day warm enough for a flight.

Cellar Wintering.—Cellar wintering is the most generally

Fig. 117.—The packing boxes may be utilised for chicken coops in summer.

practised plan by extensive honey producers of the northeastern States. Over large areas there are long periods that the bees are unable to fly from early December until the last of February or even March. It is the usual practice to put the bees into the cellar in this climate. The saving in stores will be considerable. In general it is estimated that not more than twelve to fifteen pounds of honey will be consumed by a colony in the cellar, though as much more should be present in the hive to insure a plentiful supply for spring brood rearing after the colony is placed upon the summer stands, as it is rather difficult to practise feeding satisfactorily at this season of the year.

Where cellar wintering is practised the bees should be taken in as soon as possible after the weather becomes so cold that there is little chance of further flight. It is well, also, to leave them

Fig. 118.—Concrete cellar for wintering.

in the cellar in spring until warm days are the rule. Many beekeepers take them out when the maples bloom. If conditions are favorable they will get honey and pollen at once and will, perhaps, be self-supporting from the time they are taken out. However, it so frequently happens that they will be unable to get anything from the field for weeks at a time in early spring that the wise bee-man should always see that they have enough old stores for emergencies. As previously mentioned, stores of good quality are essential to successful wintering in the cellar.

Essentials of a Good Cellar.—The greatest importance is attached to even temperature in cellar wintering. If the temperature cannot be controlled effectively, the bees are better off outside. Disastrous results are the rule in cellars where the temperature is up with every warm day and down when the outside temperature becomes cold. There is much to be learned, as yet, concerning conditions of wintering in general and especially

Fig. 119.—Cellar for wintering under the workshop.

cellar wintering. It is hoped and expected that the extensive experiments now being carried on by Dr. Phillips in the government laboratory at Washington will shortly solve some of these perplexing questions.

The majority of writers give a temperature of 45° F., as the ideal cellar temperature. No good reason has as yet been given as to just why this particular temperature is better than a higher or lower one. It is generally agreed that the temperature should not drop below 40°, especially if the cellar is damp, as a combination of dampness and a cold cellar result fatally for the bees; 50° is probably better than either.

Ventilation also seems to be essential, especially to rid the cellar of the surplus moisture. Good results are frequently reported from cellars closed up tight if the walls are porous and permit the escape of moisture readily. In general an even temperature and a dry cellar are supposed to be best. It is quite possible that it will eventually be demonstrated that a temperature somewhat above the regulation 45° is better if other conditions are satisfactory. At present without a basis on which to state positively we can accept the conditions generally agreed upon as best. (See Figs. 118 and 119.)

In this connection we can do no better than to describe a cellar which gives uniformly good results and in which the owner has never lost a colony that went into winter quarters in normal condition. The cellar is that used by Mr. Snyder, of the Iowa Bee-keeper's Association, who describes it as follows:

The cellar was constructed especially for the purpose and is under the shop and honey house and large enough to accommodate 200 colonies as he stores them.

First a stone wall about sixteen inches through was built. This wall was lined with hollow tile on which a coat of common plaster was applied. The cellar is ventilated by a chimney built from the ground and with an opening at the bottom and also at the ceiling. The chimney extends through the ceiling to the usual height above the roof. In addition to the chimney ventilator which is in the center of one end of the cellar, there are two three-inch ventilators in the corners at the opposite end. This supplies sufficient ventilation for cold weather. In mild weather the door of the bee cellar is left open. This opens into another cellar room used for storage purposes, all being kept in total darkness.

The bottom of the cellar is tile drained, the tile having outlet in the creek about a quarter of a mile distant. However, there is no direct outlet from the cellar, the tile being laid about three inches below the surface.

The ceiling is constructed of eight-inch joist covered with tar building paper and overlaid with patent metal lath on which a coat of plaster is applied. Overhead of course there is the floor of the workshop.

Most bee-keepers favor brick walls as they are dryer than cement or stone as a rule. If too many colonies are placed in a cellar for the size of the space available, there is a tendency for the temperature to rise in spring, from the heat generated in the hives. Crowding in the cellar should be avoided, but when such a condition becomes apparent it can sometimes be relieved by placing a piece of ice over the cluster on each hive. Some bee-keepers use a sprinkling can and sprinkle the fronts of the hives, permitting the water to run into the hive. While something of this kind may be necessary to quiet the bees, it is much better, if possible, to avoid the conditions that cause them to become restless.

Removing Covers or Bottoms.—It is a common practice to remove either the cover or bottom board from the hives as they are placed in the cellar. Some bee-keepers remove one, some remove the other, while others leave both in place with entrances wide open. It can hardly be said that there is any definite evidence as to which plan is best. A common plan is to remove the bottoms and leave the covers in place, then alternate the hives, placing one on top of the two below with a space between, thus providing ample ventilation to the cluster.

In general it would seem that where other conditions are right it makes little difference, as the bees seem to come through in good condition anyway. Where other conditions are bad the bees come through in bad shape, no matter whether top or bottom, or both, or neither be removed.

Summary.—Generally speaking, it may be said that the extensive honey producers are agreed on the following as essential to successful wintering by any plan: vigorous young queens with a large cluster of young bees, sufficient stores of good quality, and a dry situation. If the bees are wintered in cellar, an even temperature in addition is desired.

QUESTIONS

1. Why are the losses of bees in winter so heavy?
2. Discuss the essentials of successful wintering.
3. What stores bring best results and why?
4. What relation does the queen sustain to the wintering of the colony?
5. Under what conditions are paper cases suitable for winter protection?
6. Discuss the advantages of packing bees on their summer stands?
7. What are the advantages of double walled hives?
8. Describe the essential points of a good cellar for wintering purposes.

CHAPTER XIV
MARKETING THE HONEY CROP

The first essential to the successful marketing of any commodity is to have a good article put up in attractive shape. One of the worst drags on the honey market is the quantity of honey that goes to the store in propolized and unscraped sections, with travel stain, dirt, and leaking honey. It would require a customer who had a confirmed honey taste to even think of honey after looking at the article too often offered for sale. It is a fortunate thing for the business that honey production is rapidly passing into the hands of specialists who know how to prepare their product for market in attractive condition and that the small farm apiaries are rapidly passing away in most places. There is no good reason why honey in small quantities might not be as well cared for as large quantities. Too many who have but a few colonies of bees regard whatever honey is secured as that much velvet, and are satisfied to take it to the store in the easiest way possible and to accept such a credit on the grocery bill as the merchant is willing to give.

COMB HONEY

Grading.—The bee-keeper who wishes to establish a permanent market cannot place too much importance on carefully grading his product so that every package will be uniform with others of the same grade.

There is more carelessness, as yet, in the preparation of the honey crop for market than any other staple food product. For some reason the bee-keeper has not kept pace with other enterprises in the marketing of his crop, and to this he owes, to a great extent, the fact that honey does not bring as good prices as some other commodities.

Some confusion has resulted in the different grading rules

adopted by the bee-keepers of different sections. It is highly desirable that the same rules be made to apply to all sections, so that the merchant buying honey from any locality will know what to expect.

Official Grades.—The National Bee-keeper's Association at the convention in Cincinnati in 1913 adopted the following as official for the association:

Sections of comb honey are to be graded: first, as to finish; second, as to color of honey; and third, as to weight. The sections of honey in any given case are to be so nearly alike in these respects that any section shall be representative of the contents of the case.

1. *Finish.*—(1) Extra Fancy: Sections to be evenly filled, comb firmly attached to the four sides, the sections to be free from propolis or other pronounced stain, combs and cappings white, and not more than six unsealed cells on either side.

(2) Fancy: Sections to be evenly filled, comb firmly attached to the four sides, the sections free from propolis or other pronounced stain, comb and cappings white, and not more than six unsealed cells on each side, exclusive of the outside row.

(3) No. 1: Sections to be evenly filled, comb firmly attached to the four sides, the sections free from propolis or other pronounced stain, comb and cappings white to slightly off color, and not more than forty unsealed cells, exclusive of the outside row.

(4) No. 2: Comb not projecting beyond the box, attached to the sides not less than two-thirds of the way around, and not more than sixty unsealed cells, exclusive of the row next to the wood.

2. *Color.*—On the basis of color of the honey, comb honey is to be classified as: first, white; second, light amber; third, amber; and fourth, dark.

3. *Weight.*—(1) Heavy: No section designated as heavy to weigh less than fourteen ounces.

(2) Medium: No section designated as medium to weigh less than twelve ounces.

(3) Light: No section designated as light to weigh less than ten ounces.

In describing honey under these rules, three words or symbols are to be used, the first descriptive of the finish, second the color, and third the weight. For example, fancy, white, heavy ($F = W = H$). No. 1, amber, medium (No. $1 = A = M$). In this way all the combinations of color, weight, and finish can be briefly described.

Cull Honey.—Cull honey shall consist of the following: Honey packed in soiled second-hand cases or in badly stained or propolized sections; sections containing pollen, honey-dew honey, honey showing signs of granulation, poorly ripened, sour or "weeping" honey; sections with comb projecting beyond the box, or well attached to the section less than two-thirds the distance around the inner surface; sections with more than sixty unsealed cells exclusive of the row adjacent to the wood; leaking, injured, or patched up sections. See Fig. 120.

Fig. 120.—Development of comb honey in sections: 1, full sheets of foundation; 2, cells drawn and partly filled; 3, cells filled and partly capped; 4, fancy comb honey ready for the table

From the above rules it will be seen that it is a very short-sighted policy to mix inferior sections with the good ones in the hope of getting a better price for all. The result is to bring the price of the lot down to the level of the poorest grade.

The more carefully and conscientiously grading is done the better price will be obtained and the easier to find a ready sale. Good quality comb honey carefully graded will nearly always sell readily, although like other commodities the price varies with seasons.

Commission Houses.—There has been much complaint from disappointed bee-keepers who have not been satisfied with results from sales through commission houses. Sometimes the fault is with the producer and sometimes with the commission man. The bee-keeper should exercise ordinary business methods and not consign goods to a commission firm without some knowledge of the standing of the firm. While large quantities of honey are sold through commission firms to the regular trade, it often happens that so much honey is sent to the larger centers as to greatly depress the market, while the markets in the smaller places may be short. The problem of proper distribution of the honey crop is a serious one and demands careful attention on the part of the producer who would realize the most from his product.

There are commission firms which specialize in the sale of honey and which handle large quantities to the satisfaction of their clients. There are other concerns that buy outright in carload lots at an agreed price. As a rule, a cash sale with no chances, even at a slightly lower price, is to be preferred.

Home Markets.—As a rule the bee-keepers living east of the Rocky Mountain region will find it greatly to their advantage to develop home or nearby markets. The western honey producers are at a disadvantage in this respect, for bee-keeping is more highly specialized in the West and the honey produced is greatly in excess of what home markets can absorb. It thus becomes necessary to seek distant markets. In such localities co-operation is a great advantage in reaching a profitable market.

In the eastern States there are few localities where a profitable market for large quantities of honey cannot be developed within fifty to one hundred miles. Methods of developing local markets will be considered more in detail under "Advertising Methods."

Shipping Cases.—Comb honey is a perishable product and considerable care is necessary to see that it is packed in such shape that it will stand the rough handling necessary to shipment and reach its destination in presentable condition. If the average bee-keeper could spend a day in some large warehouse where large amounts of honey are being received and take note of the number of cases that arrive in broken and leaking condition he would not be surprised at many of the poor returns from shipments. A single broken section may spoil the appearance of a whole case by dripping over the remainder which may be in otherwise good condition. So much irritation arises in making settlements for honey that arrives in bad condition that many responsible firms refuse to handle it at all. Then again dishonest firms take advantage of the fact that honey is so commonly damaged in shipment to report as damaged goods that really arrive in good condition.

Wood shipping cases holding twenty-four sections are the most common as yet. The corrugated paper in bottom of case offers the advantage of apparently greater safety to the honey and does not add greatly to the cost. The paper will absorb much of the shock of rough handling as well as part of the leakage of broken sections.

It is very desirable that comb honey sent to market should be enclosed in paper cartons. This offers several advantages. The section is kept clean and free from dust and also retains all its own leakage, thus saving the loss resulting from soiling other sections.

Corrugated paper shipping cases have also been tried to some extent but as yet have not been widely used. Their value remains as yet to be demonstrated. Fear is expressed that they will not protect the fragile contents as well as the wooden case with paper lining.

The use of the best possible protection to the honey shipped to market is cheap insurance and the risk of resulting loss will be sufficiently reduced to overbalance the greater expense.

Care of Comb Honey.—Comb honey should be fully finished and ripened before taking from the hive, but should not be left until the appearance is spoiled by travel stain.

As soon as it is removed it should be stored in a warm room. Care should be used that it does not freeze, as low temperatures hasten granulation and granulated comb honey is likely to be a "drug" on the market. While candied extracted honey can readily be liquefied, it is difficult to do anything with granulated comb honey. In this case "an ounce of prevention is worth a pound of cure." Fortunately comb honey does not usually granulate quickly and there is usually ample time to dispose of it before it will begin to candy in the comb.

The novice is likely to store his honey in the cellar, the worst possible place for it, thinking to keep it cool. The author sometimes receives letters from bee-keepers who have spoiled a nice lot of honey by storing it in a cold, damp place, wishing to know what can be done to restore it. If it is merely candied the situation is not so bad, but honey stored in a cold, damp cellar gets weepy and sour so that it is of little use for any purpose with which the author is familiar, unless it be for making vinegar.

A warm and dry place is the best storeroom for honey. It should by all means be dry. Well ripened honeys are much less likely to granulate and for this reason the honey gathered early in the season gives much less trouble than that gathered late in summer. The source from which the nectar is gathered makes some difference also, as some honeys are much more likely to candy than others subjected to the same conditions.

Early Sales.—As a rule the early market is best and the man who depends on the general market instead of establishing a market of his own which he supplies through the year, will do well to sell as early as possible after the crop is removed from the hive. Most years there is a period following the holiday season when honey moves slowly and new shipments do not receive prompt returns. As a rule, it is easier to sell early and the prices average as good or better than later. The fellow with a market established is independent of the fluctuations of the general market and the increased returns from retail sales pay well for the additional time required, in most cases.

EXTRACTED HONEY

Packages for Extracted Honey.—There is less risk of loss of extracted honey in shipping than of comb honey. Of course, an occasional can will leak and an occasional package of glass be broken but the total breakage of comb honey is several times greater than that of extracted. The producer draws his honey from the extractor into large settling tanks where it can remain safely until he is ready to place it in the packages in which it goes to market. If he depends largely on the general market the honey will mostly be shipped in square cans holding sixty pounds each. Two of these cans are shipped in one case, making a package of 120 pounds weight.

If he develops a special or retail market he will use such a package as his market demands. Usually the bee-keeper determines the kind of package in which he prefers to sell his product and educates his customers accordingly. It takes several years to develop an extensive retail trade but it is a very satisfactory way to dispose of one's crop and is not difficult to do (Fig. 121).

The friction top pail holding five pounds is a popular package for retail trade. Five pounds is not too much for the average family to buy at one time, although many buy in smaller packages. For the grocery trade jars of eighteen ounces, one and one-half pounds and three pounds capacity are perhaps the three most popular sizes. For the grocery trade glass containers are much to be preferred to tin, as the contents can be seen which adds much to the total sales.

A one-quart fruit jar of white clear glass makes a very satisfactory package for retail trade. It holds about three pounds net. A tin cover with oiled paper lining should be used instead of the usual top and rubber used for canning fruit.

Controlling Prices.—One of the most satisfactory kinds of trade to develop is a fancy retail trade through regular grocery stores. If the bee-keeper will guarantee to make good any losses from any cause and to take back goods that are unsatis-

Fig. 121.—Packages for retailing extracted honey.

factory for any reason without quibbling, he can frequently put in his line in the best groceries and set the retail prices, allowing the grocer about twenty per cent of the selling price. If care is used to offer only a high grade and uniform product a select trade can be established which will demand a certain brand and which in a few years' time will be a good asset to the bee-keeper's business.

There are so many different grades of honey from different sources that the average consumer becomes confused when different purchases are of such decidedly different quality and comes to fear adulteration. It is very important that a certain trade be supplied with a uniform quality, as this is one of the best ways to hold trade.

Blending.—If the bee-keeper has honey from several sources so that his product varies greatly in quality and flavor he should either blend the different kinds together as a whole so as always to have his brand uniform, or he should use only his best honey under his brand and dispose of the other stock on the general market. Blended honey gives very good satisfaction, usually, if the blend is always alike. This plan permits the bee-keeper to dispose of all his product in his own trade and it brings better prices.

If one has poor quality honey of any kind he should not take chances of spoiling his market by using it unless it is his principal source, in which event he can develop a market which will come to demand that particular product.

Co-operative Marketing.—Where the business of honey production is highly developed, as in some sections of the West, the co-operative plan offers decided advantages. Many of the bee-keepers are engaged in production on such an extensive scale that they find little time or inclination to bother with the selling end of the business. If the coöperative association is in the hands of competent managers the honey goes to the best markets and the large volume of business transacted cuts the cost of handling down to the lowest possible figure.

Fig. 122.—Trade-mark of the Colorado Honey Producers Association.

The Colorado Honey Producers' Association is one of the most successful of these coöperative associations. The individual member packs and marks his honey according to the association rules and ships it to the Denver headquarters. If he has a sufficient quantity to ship it out in carlots the manager of the association or someone for him inspects the honey to see that it is properly graded and it is shipped to market directly from the apiary. The bee-keeper gets the full amount of cash resulting

Fig. 123.—Honey label.

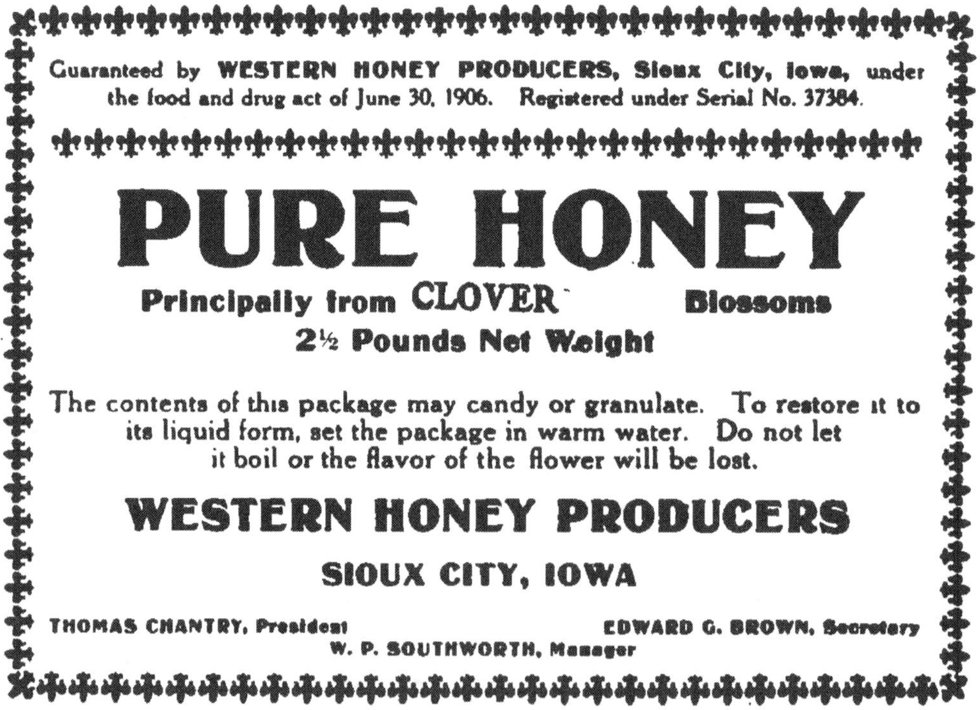

Fig. 124.—Honey label.

from the sale, less commission and expenses, in a few days, so that as far as he is individually concerned it is a cash transaction. Where the producer must seek distant markets this plan offers the maximum of return possible with a minimum of trouble. It is the association, instead of the individual, that looks after such details as correspondence, collections, shipments, etc.

Under this plan great care is used to have all grades of both

Fig. 125.—Honey label.

comb and extracted honey of uniform quality and the association brand soon comes to be known in the markets. Figure 122 shows the trade mark or brand adopted by the Colorado Honey Producers' Association. Figure 123 shows their label for large packages. (See also Figs. 124 and 125.)

ADVERTISING METHODS

In the development of special or retail markets suitable advertising is of the greatest value. This subject can well be considered from two angles, that of general advertising which has for its object to increase the consumption of honey, and advertising the product of a particular apiary for the purpose of establishing a direct-to-consumer trade.

Methods of General Advertising.—There is not a great deal that the individual bee-keeper can do in the way of general advertising, because the expense is prohibitive. Dr. Bonney's little red stickers (Fig. 127) are as good as anything yet proposed. These little stickers are printed and offered for sale by several enterprising firms at thirty-five cents per thousand and their use has become general among the bee-keepers almost in a day. Thousands of them are pasted on envelopes containing outgoing mail, and in all kinds of places where they are likely to attract the attention of the public. One of these little stickers attached to a letter will attract the notice of several carriers and clerks in the postal service before finally being delivered to the person to whom it is addressed.

Fig. 126.—Honey label.

Dr. Bonney has found some new customers among the mail clerks who have been attracted to the return card of "Bonney Honey, Buck Grove, Iowa," on the envelopes he uses in his correspondence. He also uses a sign at his apiary as shown in Fig. 128.

The Iowa Bee-keeper's Association has adopted rather a novel plan of general advertising at the holiday season. A large placard is printed in two colors, with the words, "Eat Honey with Your Christmas Dinner" (Fig. 129), and the Greetings of the Iowa Bee-keeper's Association. The association has furnished to each of its members as many of these cards as he wished to place in the stores where his honey was on sale. At this season of the year when luxuries are in special demand it is quite possible to make many new customers for honey who have regarded it as a luxury not for general use. These cards attract instant attention to the honey on sale at the precise moment when the purchaser is prepared to buy something for his table, and if the packages are attractively displayed increased sales will be the result.

If the bee-keeper has a bent for advertising it would be quite possible to adapt this idea to his individual use and by preparing a series of such cards suitable for every season of the year and keeping each kind on display but a few days at a time he can add considerably to the demand for honey in the stores where it is on sale.

Fig. 127.—Little stickers widely used for general advertising.

Exhibits at Fairs.—A good exhibit at either State or county fair is not only good general advertising but also likely to be of great help to the individual bee-keeper who makes the exhibit (Fig. 130). Multitudes of people pass by such an exhibit daily and if there be a well-informed attendant he can do much to create a demand for his product on the part of the visitors.

One year the prize winning exhibit at the Iowa State fair carried off about two hundred dollars in premiums and in addition the owner took orders for about five thousand pounds of honey at retail prices. He was thus amply repaid for all the time and labor necessary to make a creditable showing for the industry in general and for his apiary in particular.

It is quite probable that half of the honey sold as a result of this exhibit was to customers who would not have gone to the store to leave an order for it.

The author once put up a small exhibit for a store where his honey was on sale, to be used at a county fair. Extracted honey in jars of the various sizes in which it was regularly offered for sale was the principal part of the exhibit. It was arranged as attractively as possible about an observation hive containing a frame of brood and live bees. Above the exhibit was placed

FIG. 128.—Advertising sign at the Bonney apiary.

a large sign painted in two colors and worded substantially as follows:

PURE EXTRACTED HONEY
Gathered by Cass County Bees
From Cass County Flowers
Expressly for Stier Grocery Co.,
Atlantic.

There was no attendant in charge of the exhibit, but according to the man in charge of the general department in which it was placed, there was more interest in it than in any other in the department. The live bees were of course the principal object of interest. Had there been an attendant in charge to answer questions and take orders the results might have been even greater, but with no word concerning its origin excepting the sign the results were a great surprise both to the grocer and the bee-keeper. Orders for honey began coming in immediately and by the close of the fair the supply available at the store was all sold and the delivery wagon sent to the fair grounds to take down the exhibit to supply pressing orders. A

EAT HONEY
WITH YOUR
CHRISTMAS
DINNER

GREETINGS
Iowa Bee-Keepers Association

FIG. 129.—Iowa Bee-keepers Association holiday placard.

hurry-up call was sent to the apiary for more honey, which was supplied at once. As a result of this single little exhibit and sign at a county fair, which did not require much more than a half day's time to prepare and put in place, the sales of honey from this store were more than doubled and many of the customers who first bought as a result of it, remained as permanent customers of this particular store and particular brand of honey as long as it remained on the market.

As a rule the bee-keeper who seeks the shortest and most

Fig. 130.—An exhibit at the fair is a good advertising medium and promotes the use of honey.

FANCY PACKAGES

direct way to reach his customers, after once trying the plan of exhibiting, remains as a permanent exhibitor as long as he cares to develop this kind of market. One will find it difficult to show his product and explain his methods to as many people in any other way. The premiums offered are usually sufficient to pay the expenses incurred by the exhibitor.

Fancy Packages.—One of the very best advertisements for the honey producer is an attractive package decidedly different from those for sale in the general market. The use of paper

FIG. 131.—Paper carton the best retail package for section honey.
FIG. 132.—The Hunten tin package.

cartons for comb honey offers good opportunities for creating a demand for a particular brand. Instead of making use of the regular stock carton with the simple addition of the producer's name it is much better to have a special design with a particular brand and have the design copyrighted (Fig. 131).

One of the most attractive packages for comb honey ever placed on the market was put out by Paul Hunten, of Colorado. Mr. Hunten had a section made of tin instead of the usual wood. When this was filled he had a top and bottom to slip on like the lid to a tin can and a paper band to go clear around the four sides and make the package dust proof. The transparent center of the one side gives a view of the contents (Fig. 132). This package would be suited only to the highest class of trade because of the extra expense to produce it, but there is a trade that would gladly pay a few cents extra for each section in order to secure a fancy package that is dust and drip proof.

The paper carton serves the same general purpose and is much cheaper. Extracted honey also sells much better in attractive packages, as any bee-keeper of experience has learned by experience. The experiment has been tried of putting honey in an ordinary Mason fruit jar with ordinary top and rubber beside containers the same quantity and quality of honey but of a clear white glass and nice fitting top and attractive label. From six to ten times as many jars of the more attractive appearing lot were sold as of the other, thus proving how far the appeal to the eye will assist in making a sale.

Retail Prices.—Many bee-keepers prefer to dump the whole crop on the general market to sell for what it will bring rather than to go to the trouble of developing the retail market. As a rule extracted honey of good quality will not sell readily at more than $7\frac{1}{2}$ to 8 cents per pound in large quantities at wholesale prices. At the same time extracted honey of similar quality will bring from ten to fifteen cents per pound net at retail with an average of about twelve cents per pound perhaps. Unless the producer has a very large business that occupies his time fully he can well afford to spend considerable time in marketing his product for the extra fifty per cent.

The small apiary that produces from $1000 to $1200 per year can thus be made to pay from $1500 to $1800 annually. While to make the most of such a market will require that honey be kept in stock to supply the trade throughout the entire year, most of the additional work will be required during the months when least is required in the apiary. There is the further advantage that every man who develops his own market relieves the general market to that extent and thus serves to steady prices or even to advance them.

Newspaper Advertising.—Direct advertising offers a very good field if the copy is well arranged and the best medium selected. Too many producers confine their advertising to the bee journals. These are read principally by other producers and the only buyers are bee-keepers who have a larger market than they can supply, but they buy only at wholesale prices or little above.

The buyers which can be reached profitably are the real consumers and especially those who buy in considerable quantity. Western farmers and ranchmen are good customers, especially in sections that are a long distance from the railroad and where supplies must be purchased long in advance. Some of these ranches will buy as much as half a ton of extracted honey at a single order. The farm and ranch journals that circulate in the arid regions where ranching is still carried on extensively furnish good advertising mediums for the sale of honey. The farm journals which circulate in the Mississippi valley are also good mediums, as the farmers of the Middle West are prosperous and less honey is produced by the general farmer every year.

Local newspapers can usually be used to advantage. In making use of the local paper the producer can offer to deliver his product on telephone order. Much depends upon the wording of the advertisement, no matter what medium is used. The mere mention of honey for sale at a stated price will bring orders from customers who are already consumers of this product, but will seldom attract the attention of others. An advertisement with some novel suggestion will attract the attention of the casual reader and often bring an order.

HONEY THAT TASTES LIKE MORE

Our new honey is now ready for delivery. The bees have been unusually busy this summer and the product is of the finest quality. Flowers are nature's supreme effort and honey is the essence of the flowers. A sample of our clover blend will convince you that a finer food product has never been produced. Only fifteen cents per pound in ten pound lots.

CLOVERDALE APIARIES

In "Advanced Bee Culture" W. Z. Hutchinson gives an account of an advertising experience by which he sold ten thousand pounds of honey from a single advertisement in *Saturday Evening Post* at a cost of $25. The magazines of national circulation offer a field of their own which the ordinary bee-keeper is hardly prepared to cultivate. The circulation is so widely scattered and the cost is such that there is little hope that advertising in this way will prove profitable unless the bee-keeper has attractive printed matter which he is prepared to send in answer to every inquiry together with a sample of the honey.

A large producer who is prepared to follow up inquiries and who has well prepared printed matter giving some information as to the production of honey and its preparation for market may find advertising in these high class journals profitable. As a rule, the novice should begin with his local papers, then gradually increase his advertising appropriation as he learns how to make the most of it.

The local market can always be most profitably developed and in most localities east of the Missouri River the bee-keeper need not seek the distant market.

Booklets.—No matter what method one may take to find his customers a cheap booklet giving the uses to which honey can be put will be of great value. This should be printed on good paper with some attractive pictures of apiary scenes and honey packages. There should be information concerning the care of honey. Too many people will take home a section or two of honey and spoil it by putting it in the refrigerator. The manner of liquefying granulated honey should always be given.

This should be followed with some brief descriptions of the methods of honey production and preparation for market, and a number of receipts for the use of honey in cooking or other household uses should be included. One of the best things of this kind is the 54-page booklet, "The Use of Honey in Cooking," published by the A. I. Root Co., and distributed for a short time by the National Bee-Keeper's Association. These booklets are published in such large quantities for sale to honey producers for advertising purposes that they can be purchased at much less than a similar book could be printed for. These are now sold at about $4.00 per 100. By putting one of these or some similar matter in the hands of each purchaser of honey the demand is likely to be stimulated for years after, as a result.

Shortcake.—Three cups flour, two teaspoonfuls baking powder, a teaspoonful of salt, ½ cup shortening, 1½ cups sweet milk. Roll quickly and bake in hot oven. When done, split the cake and spread the lower half thinly with butter and the upper half with a half pound of the best flavored honey. (Candied honey is preferred. If too hard to spread well it should be slightly warmed or creamed with a knife.) Let it stand a few minutes and the honey will melt gradually and the flavor will permeate all through the cake. To be eaten with milk.

Soft Cake.—One cup butter, 2 cups honey, 2 eggs, one cup sour milk, 2 teaspoonfuls soda, a teaspoonful each of ginger and cinnamon, four cups flour.

Eggless Cake.—One cupful sugar, ½ cup honey, one cupful sour milk 2 tablespoonfuls butter, one cupful chopped raisins, one cup chopped dates, 1 teaspoonful soda, 2½ cups flour, spice to taste.

Gingerbread.—One egg, one cup honey, one cup sour milk, 2 teaspoonfuls butter, ½ teaspoonful soda, one teaspoonful ginger. Flour to make rather stiff batter.

Honey Jumbles.—Two quarts flour, 3 tablespoonfuls melted lard, one pint honey, ¼ pint molasses, 1½ level teaspoonfuls soda, a level teaspoonful salt, ¼ pint water, ½ teaspoonful vanilla.

Ginger Cookies.—One cup each of honey, sugar, buttermilk and lard; one teaspoonful each of salt, cinnamon, and ginger; one teaspoonful soda; one teaspoonful lemon extract. Stir stiff with flour, for gingerbread; mix stiff and roll and cut and bake in a quick oven. Also good with caraway seeds instead of spices.

Oatmeal Cookies.—Cream together one cup sugar, ⅓ cup honey, ¾ cup lard or butter, 6 tablespoonfuls milk, ½ cup raisins, 2 cups rolled oats, 2 eggs. Sift together 2 or more cups flour, ½ teaspoonful salt, 2 teaspoonfuls cream tartar, one teaspoonful each of soda and cinnamon. Mix together and roll quite thick.

German Christmas Cookies.—One quart honey. Let it come to a boil, then set away to cool. Add one pound brown sugar, 4 eggs, juice and rind of two lemons, ¼ pound citron chopped fine, 2 teaspoonfuls soda, one tablespoonful each of cinnamon, cloves, allspice and nutmeg. Flour to stiffen. Make dough as stiff as you can. Chopped nut meats may be added if desired.

Brown Bread.—One cup corn meal, one cup rye meal, one cup sour milk, ½ cup or less of honey, teaspoonful salt and teaspoonful of soda. Steam four hours and then dry in oven fifteen minutes.

Graham Bread.—Take 1½ cups sour milk, ½ cup shortening, ⅔ cup honey, one egg, teaspoonful soda, 3 cups graham flour.

Bran Gems.—Two cups bran, one scant cup wheat flour, one pinch salt, 1½ cups buttermilk, level teaspoonful soda, 3 tablespoonfuls extracted honey. Mix the bran and flour and salt thoroughly; add buttermilk in which soda has been dissolved; lastly add honey. Bake until thoroughly done, in greased gem-pan in hot oven.

Sandwiches.—For an afternoon tea or lunch cut thin slices of bread and spread with honey quite thick. Use brown or whole wheat bread, or use one kind of bread for top layer and another kind for bottom. For a richer sandwich sprinkle with nut meats or sugar.

Honey Cereal Coffee.—One egg, one cup honey (preferably dark), 2 quarts wheat bran. Beat the egg, add honey and lastly the bran, and stir until well blended. Put in oven and brown to dark brown, stirring frequently, being careful that the oven is not too hot. To prepare the coffee, allow one heaping tablespoonful of the brown mixture to a cup of hot water, and boil for at least ten minutes. If properly prepared this is equal or superior to any cereal drink on the market.

Apple Butter.—One gallon good cooking apples, one quart honey, one quart honey vinegar, one heaping teaspoonful ground cinnamon. Cook several hours, stirring often to prevent burning. If the vinegar is very strong use part water.

Peach Preserves.—Pare and halve nice large peaches the night before. Pour one pound of honey to every one and a half pounds of fruit.

Honey Crab Apple Jelly.—Boil fruit with as little water as possible; squeeze through jelly bag. Add one-half cup honey and one-half cup of sugar to each cup of juice, then boil twenty minutes or until it begins to jell. Pour into glasses to cool but do not cover until fully cooled.

Baked Apples.—Split some sour apples, cut out the cores and fill pan. Bake until they begin to soften, then fill cavities with honey and lemon juice. Set back in the stove to finish baking.

Honey Candy.—Take 2½ cups sugar, ½ cup honey, ½ cup water and boil to thick syrup. Pour one cup of syrup on beaten whites of two eggs, stirring meanwhile. Boil remainder of syrup until it hardens when dropped in water; then pour in the syrup and eggs, stirring briskly. Add a cupful of peanuts and stir until it begins to harden, then spread in a pan and cut in squares. Flavor to taste. If properly made it will be soft and pliable.

Honey Pop-corn.—Take a teacupful of white honey, a teacupful of white sugar, 1½ tablespoonfuls butter, a tablespoonful water and boil until brittle when dropped in cold water. Have ready two quarts nicely popped corn and pour the candy over it until evenly distributed, stirring briskly until nearly cool.

Candy.—One cup granulated sugar, one tablespoonful extracted honey, butter the size of a walnut, and sweet cream enough to dissolve the mixture. It needs but little cooking. When taken from the fire beat until smooth.

Candy.—One cup sugar, 2 tablespoonfuls honey, 2 tablespoonfuls water and walnut meats. Cook and test like molasses candy.

Taffy.—Take three cups sugar, ⅔ cup extracted honey, ⅔ cup of hot water. Boil all together until it spins a thread when dropped from a spoon, or hardens when dropped into cold water. Pour into a greased pan to cool, when it should be pulled until white.

Fig. 133.—Dr. Bonney's postcard which brings him many new customers.

The Use of Post Cards.—One of the most effective means of advertising in a small way is the use of post cards. Dr. A. F. Bonney, of Iowa, has used this method quite extensively. The post cards mention honey only incidentally but are usually somewhat comic in makeup. Fig. 133 shows one of the cards which he has used to a considerable extent. His plan is to send them to postmasters, public officers, and prominent and prosperous people generally whose names he can secure within one hundred miles of his home. It would be well to use two or three lines at the bottom of such a card as that here shown to quote prices of honey delivered in packages of popular size.

The idea of these cards is to catch the interest of the recipient who will laugh at the comic picture and then have his attention called to the honey which is unobtrusively done. One of Dr. Bonney's cards pictures the occupants of an automobile in all sorts of impossible situations as the result of an accident. Nailed to a tree in the background is a sign board with these words: "If anything happens in the vicinity of Buck Grove, Iowa, stop and get some Bonney Honey."

When put to the test of practical results they have proved to be good business getters. After sending out a batch of these cards, even though they go to entire strangers with whom he has had no previous correspondence, he always gets a bunch of orders as a result.

Canvassing and Peddling.—This method is distasteful to many bee-keepers yet it has decided advantages over other plans. If one is adapted to canvassing he can take a can of honey for samples and by making a house to house canvas make many permanent customers. By offering a sample of his product the buyer is given a chance to decide whether the flavor appeals to his particular taste. Then the producer can give some information concerning the production of honey and correct any false impressions concerning the product of the hive. A good canvasser will make good wages over and above wholesale prices even if the value of future orders is not considered. A large producer can well afford to hire students during the vacation period and put them to work in building up a trade. Of course it will be necessary to sift the possible applicants somewhat to find those who are adapted to canvassing and also who know enough about the bee-keeper's business to answer questions intelligently.

A more common method is for the producer or his agent to take a spring wagon or auto with a load of honey and deliver his orders as he goes (Fig. 134). A good salesman will sell

FIG. 134.—The automobile as a sales agency is the most up-to-date method.

several hundred pounds daily from a wagon. One of the most successful honey producers of the Middle West takes his load of honey and visits the public sales that are held within reach of his home. At the public sale a considerable crowd is always gathered and he has a good opportunity to dispose of his wares to advantage. In this way he sells a good many thousand pounds during the winter months when sales are in progress. By driving ten or twelve miles in every direction he is thus able to cover a large territory and present the merits of his product to many hundreds of men. He carries packages both large and small and is prepared to supply any desired amount from a five pound pail to a sixty pound can on the spot. In this manner he sells at times more than a ton of honey within a week. If he gets but two cents per pound more than it would bring at wholesale he is making good wages for his time while establishing a trade that will soon come to depend upon his apiary as a source of supply.

Cutting Prices.—One of the worst drawbacks to the honey business is the tendency on the part of some to cut prices. John Smith will make enquiry of some concern dealing in honey as to the price they are paying. They will of course quote a price at which they can handle his goods at a profit. Mr. Smith thinking to accommodate his neighbors sells his honey at home at the wholesale price. When the supply is exhausted there is bitter complaint against paying at retail more than the wholesale price. The dealer of course must feel that he paid too much for the crop and accordingly he starts in the following year to buy at a lower figure. The retailer's profit is as legitimate as the producer's profit. Unless a man will sell at a fair retail price he should in justice to other bee-keepers if not to his own future prosperity sell it to some dealer at wholesale. Cutting prices can have but one result: the tendency to depress prices below the point of profitable production. Until the bee-keepers of a community come to practice good business methods in handling their crops the business of honey production will not be a profitable one nor will the public regard it as a desirable occupation.

QUESTIONS

1. Why is it fortunate that bee-keeping is becoming a specialized business?
2. Discuss the grading of comb honey.
3. How can the crop be marketed to the best advantage?
4. What precautions are necessary in caring for honey?
5. Note the best packages both for shipping and for retailing.
6. What plans can be used to develop local markets?
7. Under what circumstances is coöperative marketing desirable?
8. Outline some practical plans of advertising the product of the apiary.
9. What is to be expected from a general exhibit as an advertising medium?
10. Discuss the value of an attractive package.
11. To what extent can newspapers and booklets be profitably used?
12. When is personal canvassing profitable?

CHAPTER XV

LAWS THAT CONCERN THE BEE-KEEPER

BECAUSE of the nature of the honey-bee and the fact that the insects cannot be restrained like cattle or poultry, the laws concerning the bee-keeper are somewhat different from those that affect the owners of other live stock. In the first place bees are recognized as being wild by nature and once a swarm gets beyond its owner's control and passes to the premises of another he loses all property right in them unless he follows them and keeps them in sight.

Bees found in a tree or other natural cavity become the property of the first person who reclaims them. This fact, however, does not give any right to trespass on the property of another. During the days of early settlement of this country there was an unwritten law that wild bees became the property of the man who found them and marked the tree. While this right was generally recognized there was no law that would confer any right to the bees unless the finder proceeded to take possession of them. As soon as wild bees are taken into possession they become the property of the man who reclaims them. This right will be recognized and protected as long as they are under his care. Should he injure the tree in which the bees are found in removing them, he will be liable to the land owner for trespass.

The time has gone by, in most localities, when serious questions regarding the ownership of wild bees are likely to arise. Bee-keeping is now a recognized industry in itself and the owner of bees enjoys the same rights and privileges as holders of other property. The relation of the bee-keeper to his neighbors, however, especially where there is a large apiary in close proximity to the home of other persons, frequently presents some problems that are decidedly different from those of any other calling.

The keeping of bees in cities and towns is so generally practised and has been the source of so much litigation of one kind and another that an extended account of the rights of both the bee-keeper and his neighbor can very properly be taken up. While the courts have held that bee-keeping is a legitimate pursuit and as such cannot be prevented by general legislation that declares the bees to be a nuisance whether they are so in fact or not there is a general principle that will provide relief from undue annoyance.

Causes of Trouble.—Before taking up the consideration of the law in this special relationship it may be well to consider the causes that lead to friction between the bee-keeper and his neighbors. So many instances of trouble of this kind arise that small towns and cities are frequently urged to pass ordinances to prohibit the keeping of bees within the incorporated limits.

Spotting Clothes.—When the bees are brought from the cellar in spring or when they are able to take their first flight after long confinement the abdomens are distended with retained faeces. As soon as they can fly this is voided in large drops of offensive refuse. If it happens that the bees fly for the first time when the wash is on the line the white clothes are badly soiled as a result.

The bee-keeper should avoid if possible setting cellar wintered bees out when the neighbors are washing. Bees seldom fly far on the first flight and clothes are not likely to be soiled far from the bee hives. As a rule it is the near neighbors who will be the sufferers. If the bees are likely to fly on wash day the situation should be explained and some provision made to avoid having the clothes exposed. After two or three nice days there will be little further trouble, as this spotting is only noticeable after long confinement without opportunity to void the excrement.

Watering Places.—With bee-keepers as with others "An ounce of prevention is worth a pound of cure," and if the bee-keeper is diplomatic he can usually avoid annoying his neighbors seriously. No fair minded man will wish to annoy others, whether or not he is living within his legal rights.

Watering places where the bees congregate in large numbers are frequently sources of great annoyance, as animals that come to drink are likely to be stung as well as persons whose duties take them there. After the bees come to frequent such a place it is a little difficult to check their coming unless the water can be covered in such a way that they cannot reach it.

The wise bee-keeper will provide watering places for his bees as described in Chapter IV, early in spring to prevent, as far as possible, their going to other places for water.

After the bees have formed the habit of getting water at places where their presence is annoying the bee-keeper should assist in every possible way to cover the water supply until they begin going elsewhere.

Flying about Streets or Highways.—If the hives are situated near the street or highway in such a way that the entrance of the hive faces the thoroughfare there is danger of passing teams or pedestrians being stung. The bee-keeper should see that his hives are so placed that the bees do not fly directly into any public highway. The entrances should face in the opposite direction and if necessary a high board fence or other obstruction should compel them to rise high in the air before crossing. This will carry them safely over the heads of passers-by. Where persons or animals are injured by bees under circumstances such as these the owner has been held liable for damages.

In Quebec there is a legal regulation that requires that where apiaries are within thirty feet of a house or public road a board fence at least eight feet high must be erected and the fence must extend at least fifteen feet beyond the limits of the apiary. According to the editor of the *American Bee Journal* the bee-keepers feel that this regulation is a protection of their interests since they may keep bees anywhere by complying with the law.

At Candy Stores, Etc.—It frequently happens that bees will be troublesome where candy is exposed for sale or where the housewife is canning fruit or making jelly or anything else where sugar is used in making syrup. Where the doors of grocery stores are left open the bees are also likely to find some attraction.

Such annoyances as the above described are usual only during warm weather where there is no natural source of supply. After the honey flow is checked the bees are very persistent in hunting for everything sweet. Seldom is the bee-keeper to be blamed in cases like these. If the premises are properly screened against flies the bees will be unable to enter.

In Adjoining Fields.—It frequently happens that the bee-keeper will have his bees near the fence and that they will annoy the owner who cultivates the adjoining field. It devolves upon the bee-keeper to do what he can to relieve the situation by erecting a suitable fence, moving the bees, or whatever remedy may be reasonable.

While the bee-keeper has the same right to conduct his business as any other man enjoys, he must recognize the right of the public to be kept free from undue annoyance. With the foregoing causes of trouble in mind the reader will appreciate the following able discussion by J. D. Gustin, an attorney of Kansas City, Missouri, whose statements may be regarded as authoritative.

BEES AS A NUISANCE

Increasing population, greater dissemination of knowledge, and the development and specialization of industries, pursuits, and occupations combine to add constantly to the complexity of the relations of individuals, and to call from time to time, for the readjustment of the affairs of men to meet changed and changing conditions. In no other branch of the law is the ingenuity of the courts more heavily taxed in this manner than in the subject of nuisances, where, from the very nature of the subject, first principles, rather than specific legislative enactment, must always exert a controlling influence. The lawmaking power may, as occasion seems to require, declare that particular objects, actions, omissions, etc., shall be nuisances, either with or without regard to attending conditions or circumstances, but the application of such statutes is necessarily so limited that the general law of the subject is not affected.

---

[1] Bees as a Nuisance, Third Annual Report of the State Inspector of Bees, Iowa, 1914.

It therefore follows that courts still deal with nuisances largely from the principles of the common law and it is a matter of serious doubt whether, in any instance, specific legislative action can be proven to have any substantial value as an addition to the law of the subject. A nuisance at common law is that class of wrongs that arise from unreasonable, unwarrantable, or unlawful use by a person of his own property, real or personal, or from his own improper, indecent or unlawful personal conduct working an obstruction of or injury to a right of another, or of the public, and producing such material annoyance, inconvenience, discomfort, or hurt that the law will presume a consequent damage.

Text writers and legislative enactments state many variations of the foregoing comprehensive definition from Mr. Wood's treatise on nuisances, but there is no substantial disagreement as to what constitutes a nuisance. Another definition stated broadly as a general proposition, is that every enjoyment by one of his own property which violates in an essential degree the rights of another is a nuisance; and this substantial violation of a right is the true test of a nuisance, for it is not every use of his property by one which works injury to the property of another that constitutes a nuisance. Injury and damage are essential elements of a nuisance, but they may both exist as a result of an act or thing which is not a nuisance, because no right is violated. On the other hand, the pecuniary injury may be insignificant and the act or thing causing them be such an invasion of the rights of another, or of the public, as to constitute a nuisance for which an action for damages or for abatement will lie.

Nuisances are classified by the law as public and private, and there is authority for a third class called "mixed" nuisances. A nuisance is public where it affects the rights of individuals as a part of the public, or the common rights of all the community alike; a private nuisance is one affecting a single individual, or individuals of a particular class, group, or locality in a private right; the third class, referred to as mixed nuisances, are public in their nature, but at the same time specially injurious or detrimental to one or more individuals in particular who suffer a different or greater hurt than the community in general.

Nuisances are further divided into nuisances *per se*, or such as are declared so by the common law or by some statute, without regard to locality, surroundings, or circumstances, and nuisances *per accidens*, or those owing their hurtful consequences to some particular attendant circumstances, surrounding, location, or condition, without which they would not be unlawful. There are other less important and rather technical distinctions not necessary to be noticed here. The foregoing preliminary and very elementary observations of the general law of nuisances are necessary to a consideration of any subject with reference to its existence as a nuisance or otherwise.

It is also a frequent statement of the law, and may be accepted as authoritative, that no lawful occupation or business is a nuisance *per se*, except it be declared so by some special enactment prohibiting certain things as objectionable to particular localities. So also the reasonableness of the use of one's property may depend upon its situation, for what might be lawful in one locality would prove intolerable in another. The use of a building in the midst of a city densely populated for a storage house for hardware would not be objectionable in the slightest degree, while the use of the same building for the storage of gunpowder or other high explosives could not be permitted.

The common law, proceeding from fixed principles of universal application, and developing from the growth of civilization, has, in each succeeding period, found ready adjustment to new subjects resulting from the widening dominion of mankind over the creatures and forces of nature, furnishing a ready remedy for every wrongful encroachment of one upon the rights of another. In the times of the early law writers bees were most generally known as they existed in their original state. Hence they were called *feræ naturæ* and classed as wild animals. A property right, or at least a qualified property right, in them could be acquired by capture which, in accord with the general rule concerning wild animals, existed so long as the captor could hold them in possession. A distinction seems always to have been made between the possession of animals ferocious and those of gentler dispositions, and it was indictable as a nuisance to permit an animal of known mischievous disposition to go at large. Bees, however, seem never to have been regarded as ferocious or as likely to do injury to persons or property, and in the far greater number of instances in which they have been the subject of judicial consideration the questions at issue have concerned the property interests in them. It is doubtful now, however, if any court would denominate them as wild animals, in view of the present general state of development of the industry of honey production and the numerous instances of State legislation designed to promote and protect the breeding and rearing of bees for that purpose. In the one or two cases decided in American jurisdictions in which the question has been presented, it has been determined, in accordance with the rule above referred to, that the keeping of bees, even in large numbers and in towns and villages, is not a nuisance *per se*.

But greater interest, perhaps, centers in the question of whether or not bees may be so kept as to constitute a private nuisance, and also whether municipal corporations, as cities and towns, may restrain or prohibit their presence within the corporate limits. In answering the first proposition, it must be borne in mind that persons who dwell in urban communities must of necessity submit to such restrictions upon their absolute liberties that the dwelling of other persons therein shall be tolerable. As it is the unreasonable or unwarrantable use of one's premises or property, otherwise lawful, that contributes an essential element of a nuisance, a first inquiry in any case would be directed to this point of reasonableness of the use or occupation, and in determining this all of the surrounding facts and circumstances would enter into the consideration. The presence of one colony at a given point might be perfectly consistent with the due observance of the rights of the owner of the next lot, while a colony stationed at another point within the same distance would be obnoxious to the law. Again, one colony at a given place might pass unnoticed, while a number of colonies at the same place would be a nuisance. The habits of the bees, the line of flight, their temper, and disposition of the colonies, either separately or when collected together in numbers, might all furnish matter of more or less weight in reaching a conclusion. So also the character of the annoyance or injury done to the complainant must be a substantial element. In the only reported case involving this question it was charged, and the court found there was proof, " that during the spring and summer months the bees so kept "—140 colonies on an adjoining city lot and within

100 feet of plaintiff's dwelling—"by defendants greatly interfered with the quiet and proper enjoyment and possession of plaintiff's premises, driving him, his servants and guests from his garden and grounds, and stinging them, interfered with the enjoyment of his home, and with his family while engaged in the performance of their domestic duties, soiling articles of clothing when exposed on his premises, and made his dwelling and premises unfit for habitation." These facts were held to constitute a nuisance, against which the plaintiff was entitled to injunction and nominal damages. These facts just recited, however, probably present an extreme case, the immediate proximity of so many colonies being, no doubt, persuasive evidence that the annoyance suffered by the plaintiff was due to the defendant's use of his premises. Greater difficulty would be experienced in reaching such a conclusion if there were no colonies stationed in the immediate vicinity, a thing entirely possible under the common belief that the insects go considerable distances for their stores.

So it may be said of bees, as of other property, that no hard and fast rule can be laid down by which to determine in advance whether the presence of bees in any given numbers or at any given point will amount to a nuisance. But, not being a nuisance of themselves, as a matter of law, and absent also any general State enactment declaring them to be such, bees will not, under any circumstances be presumed to be a nuisance, but the matter will rest in the proof adduced, with the burden upon the party alleging the affirmative. But they may, upon proof of particular facts showing all the elements necessary to the existence of a nuisance, be condemned as such, either of a private or public character, as the nature of the injury might decide.

Predicated upon the theory advanced in the beginning that courts would now, if the matter were called in question, decide that bees are domestic animals, and it having already become a matter of legislative recognition that they are subject to communicable diseases, a question arises as to the liability of the keeper of diseased bees. At common law it was an indictable offense, which has been reënacted by statute in most of the States, to take a domestic animal suffering from a communicable disease into a public place or to turn it into the highway so that the disease might be communicated to the animals of other persons. It could hardly be said to be less culpable to knowingly keep diseased bees, which, by their nature may not be restrained or confined, to spread disease to the apiaries of other owners. If to turn a horse with glanders or a sheep with footrot into the highway is a public nuisance, on the same reasoning to turn bees at large to carry communicable diseases peculiar to them to other bees ought to be an offense of the same grade.

The power of a municipal corporation, as a town or village, to restrain or prohibit within its limits the keeping of bees, or to denounce them as a nuisance, is commonly reported as a fruitful source of vexation to keepers of bees, but one case only is reported as involving a judicial determination of that particular point. And here, too, a few preliminary observations will be necessary to proper understanding of this phase of the nuisance laws. Cities, towns, and villages, as municipal corporations or public bodies, receive their powers by express grant from the legislative authority of the State, and with the exception of some unenumerated powers without which the corporate body could not exercise its essential functions as such, their powers are limited to those expressly named in the grant. This grant of power is usually contained in the general laws of the State governing cities, towns, and villages, and is called the charter power, the law or statute itself being usually known as the charter. Keeping these facts in mind will aid the unprofessional man in understanding the terms to be encountered in an examination of local laws in regard to the power of a municipal corporation to legislate upon this subject.

Every State has its own peculiar policy toward these municipal corporations, and no two are exactly the same. They all, however, follow the same general plan, with variations influenced by local conditions. As the power of the State legislature is so limited that its acts must be consistent with the constitution, so the power of a municipal corporation to make by-laws, as its ordinances or enactments are commonly known, must be in harmony with its charter, with this further distinction, that while the legislature of the State may exercise unlimited discretion in all matters not prohibited by the constitution, a municipal corporation is restricted in legislative action to those matters in which it is expressly authorzed by its charter.

It is the general rule that cities, towns, and villages have conferred upon their common councils power to declare, abate, and remove nuisances. In the case of nuisances *per se*, whether at common law or by statute, or by ordinance in those cases in which the council may declare such nuisances, the power to abate by summary action is either expressly given or exists by necessary implication. Summary abatement means arbitrary removal or destruction without judicial process. Nearly, if not quite, all city charters contain grants of power to license, regulate, and restrict all businesses, pursuits, and avocations, and also a section known commonly as a " general welfare clause," by which the corporate body is empowered generally to enact such ordinances, rules, and regulations as may be necessary to preserve the peace, safety, and health of its inhabitants and promote their general welfare. To undertake to set out the specific provisions of the charters of the municipal corporations of the various States would extend this article far beyond its intended scope.

It is a cardinal rule of the courts that all ordinances must be reasonable, and that while a city may define, classify, and enact what things or classes of things shall be nuisances, and under what conditions and circumstances such things shall be deemed nuisances, this power is subject to the limitation that it is for the courts to determine whether, in a given case, the thing so defined and denounced is a nuisance in fact, and that if the court shall resolve this point in the negative the ordinance is invalid. Under this rule, in an Arkansas case, it was held that the municipal corporation could not prohibit the keeping and rearing of bees within its limits as a nuisance regardless of whether they were so in fact or not. And this case seems to have been received as announcing the correct rule in recent text works, though the point has not been raised elsewhere in controversy.

Under the rule just stated, the power of summary abatement would not exist, even though the presence of bees in a particular part of the city should be declared objectionable, but the point would rest, as has been heretofore observed, upon the proof adduced, the burden being upon the party declaring the affirmative of the issue.

LAWS PROTECTING THE BEE-KEEPER'S PROPERTY

As has already been stated, the bee-keeper is as fully protected in the property rights in bees as in any other domestic animals. Should anyone steal a colony of bees he could be prosecuted for larceny in probably any State.

Spraying While Trees are in Bloom.—There is a greater danger to the bees, however, than ordinary theft. It is a common practice to spray fruit trees with poisonous liquids to control insect pests. The fruit growers are not always sufficiently careful as to the time when these sprays are applied and the wholesale destruction of bees sometimes results from the application of sprays while the trees are in bloom. A number of States have passed laws prohibiting the spraying of fruit trees while in bloom, for the sole purpose of protecting the bee-keeper.

The law on this subject enacted by the State of New York is representative of the laws in force in the various States. It is worded as follows:

> Any person who shall spray with, or apply in any way, poison or any poisonous substance, to fruit trees while the same are in blossom, is guilty of a misdemeanor, punishable by a fine of not less than ten dollars nor more than fifty dollars; provided, however, that nothing in this section shall prevent the directors of the experiment stations at Ithaca and Geneva from conducting experiments in the application of poison and spraying mixtures to fruit trees while in blossom.

A somewhat similar law is in force in Canada. In States where such laws have not been passed there is bitter complaint on the part of the bee-keepers that their bees are destroyed or they are compelled to move their apiaries.

Poisoning Bees.—It sometimes happens that malicious persons will put out poisoned honey or syrup for the purpose of destroying the bees. It hardly need be said that such an act does not differ materially from a legal standpoint from poisoning any other domestic animals. A few States have passed specific statutes providing fine and imprisonment for the malicious poisoning of bees.

LAWS FOR CONTROL OF BEE DISEASE

Although other animal diseases have been subject to regulation by law for many years, laws relating to bee diseases are of comparatively recent date. Wisconsin was the first State to pass foul brood laws. In the year 1897 a law was passed providing for the inspection of bees and prohibiting the sale of infected colonies or appliances. N. E. France was appointed inspector and has served continuously since that date. At present more than half of the States have laws regulating bee diseases and providing for inspection. New States are added to the list every biennial period at the meeting of the various legislatures, and apparently but a few years will elapse until every State has made some such provision. The tendency is to enact cumbersome statutes in the beginning which set out in detail the method of procedure under every condition. After being put to the test of actual service there is a tendency to modify the laws and leave something to the judgment of the inspector. To begin with most laws require that the inspector be notified by three persons of the supposed existence of foul brood in a locality before he is compelled to investigate. Under such conditions disease may become exceedingly prevalent before three persons will notify the inspector. If a single notice is sufficient a neighborhood may be cleaned up when the disease first makes its appearance and many bees, as well as much expense, be saved.

It should be borne in mind that elaborately drawn laws rather tend to restrict the work of the inspector than to enlarge his opportunities for dealing with a serious condition. If the law is greatly extended to outline the various conditions which he is supposed to meet he will be restricted to such powers and duties as are expressly granted in the statute. On the other hand if his office and duties are created and defined in a short general statute he will be free to meet such situations as arise.

The New York law has been on the statute books since 1902 and a somewhat similar law several years previous to that time.

Bee inspection in that State is carried on under direction of

the commissioner of agriculture and the inspection service has the reputation of being very effectively handled. The law is as follows:

The Prevention of Disease among Bees.—No person shall keep in his apiary any colony of bees affected with the contagious malady known as foul brood or black brood; and every bee-keeper when he becomes aware of the existence of either of such diseases among his bees, shall immediately notify the commissioner of agriculture of the existence of such disease.

Duties of the Commissioner.—The commissioner of agriculture shall immediately upon receiving notice of the existence of foul brood or black brood among the bees in any locality, send some competent person or persons to examine the apiary or apiaries reported to him as being affected, and all other apiaries in the immediate locality of the apiary or apiaries so reported; if foul brood or black brood is found to exist in them, the person or persons so sent by the commissioner of agriculture shall give the owners or caretakers of the diseased apiary or apiaries full instructions how to treat said cases. The commissioner of agriculture shall cause said apiary or apiaries to be visited from time to time as he may deem best and if, after proper treatment, the bees shall not be cured of the diseases known as foul brood or black brood then he may cause the same to be destroyed in such manner as may be necessary to prevent the spread of said diseases. For the purpose of enforcing this article, the commissioner of agriculture, his agents, employees, appointees or counsel, shall have access, ingress, and egress to all places where bees or honey or appliances used in apiaries may be, which it is believed are in any way affected with the said disease of foul brood or black brood or where it is believed any commodity is offered or exposed for sale in violation of the provisions of this article. No owner or caretaker of a diseased apiary, honey, or appliances shall sell, barter, or give away any bees, honey, or appliances from said diseased apiary, which shall expose other bees to the danger of said diseases, nor refuse to allow the said commissioner of agriculture, or the person or persons appointed by him, to inspect said apiary, honey, or appliances, or to do such things as the said commissioner of agriculture or the person or persons appointed by him shall deem necessary for the eradication of said diseases. Any person who disregards or violates any of the provisions of this section is guilty of a misdemeanor and shall be punished by a fine of not less than thirty dollars or more than one hundred dollars, or by imprisonment in the county jail for not less than one month or more than two months, or by both fine and imprisonment.

The law above quoted confers abundant authority upon the inspectors without unnecessary restrictions upon their movements. If in their judgment a second visit or even a third or fourth is necessary they are free to make it. Most laws require a second visit of the inspector whether or not it seems necessary.

Separate Departments.—Some States have a separate department for bee inspection. The officer is designated a State official and is usually appointed by the governor. The office thus becomes a political appointment and is subject to the dangers of such a system. If a competent man is placed in charge the results are likely to be satisfactory but he is never so free in the discharge of his duties as officers whose appointment depends solely upon efficiency. It frequently happens that men who know little about bees and less about foul brood will have disease in the apiary and will refuse to be convinced of its real nature. The enmity of such men is likely to be a heavy liability when the official asks for reappointment. If, perchance, the governor is a man who is more interested in his own political future than in the welfare of the State he will be slow to reappoint men who have antagonized any considerable element.

If an inspector is reasonable and diplomatic he can disarm much of the antagonism but it is impossible for any man in this work to please everybody and do his full duty.

Under State Entomologist.—In several States the State entomologist is given supervision of bee inspection. This should give better results than a political appointment, especially in those States where the entomologist is an official of the agricultural experiment station.

Department of Agricultural College.—The various States are rapidly adding bee culture to the departments of the agricultural colleges. The best results are likely to result from placing the bee inspection under direction of the head of the department of bee-keeping. His position is such that an inefficient man will not be placed in charge and the work can be organized in connection with the school in a very satisfactory manner. Massachusetts and Ontario follow this plan.

County Inspectors.—Several States have adopted the county system of inspection. In these States the county board may appoint a county bee inspector on petition of a certain prescribed number of bee-keepers. The official is thus accountable to the local officials and receives his pay from county funds. California adopted this plan many years ago and still retains it.

While good results often come through this system local influences often result in inferior service. Serious charges have been made in some cases of inspectors using the authority of the office to remove other bee-keepers from coveted territory and the destruction of healthy bees through jealousy. While the county system is better than none at all it is a general rule that police regulations are better enforced through a State or national administration than through a local one.

Colorado Plan.—In Colorado the State and county plans are combined. There is a State appropriation administered by an inspector appointed by the State entomologist. The county boards also have authority to appoint local inspectors as in California. In this State the combined forces work together with good results. The general supervision of the State inspector has a tendency to check abuses that might arise through a purely local administration of the office, while the county official has the advantage of being near at hand and able to give prompt attention to reported cases.

Restrictions of Shipment.—Several States have laws that prohibit the shipment or bringing of bees into the State without a certificate of health signed by some duly authorized inspector. The difficulty with such provisions lies in the fact that men often come from other States who are unfamiliar with the law, and bees are brought in without the knowledge of the State officials.

Burden on Common Carrier.—In Iowa the burden is placed on the common carrier by the following enactment:

Section 1. *Diseased Bees.*—It shall be unlawful for any person, firm, or corporation to bring into, or cause to be brought into the State of Iowa, any apiary or honey bees infected with foul brood or other infectious disease, or bee destroying insects.

Sec. 2. *Certificate of Health.*—No common carrier shall accept colonies of bees for delivery at Iowa points unless the said bees be accompanied by a certificate of health signed by some duly authorized State or government inspector.

Sec. 3. *Violation—Penalty.*—Any person convicted of a violation of this act shall be fined not less than twenty-five dollars nor more than one hundred dollars.

Such laws are very important but it is difficult to enforce them fully as the inspector has no means of knowing when and where bees are to be moved. Disease is frequently brought into localities that have been previously free from it, by shipment of bees in emigrant cars along with other personal effects. Railroads and express companies issue instructions to their agents frequently and every agent is notified of a provision of law of the kind adopted in Iowa, with the result that some shipments at least will be checked until properly inspected.

Shipment of Queens.—By far the largest interstate business in bees is the shipment of queens. Thousands of queens are shipped through the mail and by express. Disease has often been carried with the cages in which the queens are sent through the mails. Usually cases of this kind are traceable to the use of honey from diseased colonies for making the candy on which the queens feed enroute. Postal regulations now require that queens shall be accompanied by a certificate of health from some duly authorized inspector or by an affidavit that the candy on which they are fed was boiled for thirty minutes.

The safest plan is for the bee-keeper to place the queen in a new cage without candy, or with candy which is known to be free from disease germs, before introducing into the apiary.

RELATING TO THE ADULTERATION AND SALE OF HONEY

The pure food laws are a great boon to the honey producer. For many years all kinds of adulterations of honey were in the market. The bee-keeper found it very hard to compete with these adulterations which could be sold at a very low price. Adulteration was so common that the public came to believe that all liquid honey was adulterated and extracted honey fell in price to such a point that it was no longer profitable to produce it.

Fortunately it has never been found possible to imitate the natural product in the comb and the comb honey producer never suffered as seriously.

Stories to the effect that comb honey was manufactured at

one time were given wide circulation in the newspapers. This resulted in distrust of comb honey also. The National Beekeeper's Association and the A. I. Root Co., manufacturers of bee-keeper's supplies, offered large rewards for proof that comb honey had been successfully imitated, which helped to offset the bad effects to some extent.

Since the pure food laws have been so generally enforced there is a returning confidence on the part of the public that extracted honey may be pure and the price has advanced with the increased demand until it is now as profitable as comb honey production. Several years time will be required to overcome the bad effects of the unfortunate conditions of other days.

While the general laws of the nation and of the various States that apply to weights and labels of food products include honey, some States have passed specific laws prohibiting the adulteration or misbranding of honey.

New York Law.—The statute of New York is worded as follows:

Defining Honey.—The terms "honey," "liquid or extracted honey," "strained honey" or "pure honey," as used in this article shall mean the nectar of flowers that has been transformed by, and is the natural product of the honey-bee, taken from the honeycomb and marketed in a liquid, candied or granulated condition.

Relating to Selling a Commodity in Imitation or Semblance of Honey.—No person or persons shall sell, keep for sale, expose or offer for sale, any article or product in imitation or semblance of honey branded as "honey," "liquid or extracted honey," "strained honey," or "pure honey" which is not pure honey. No person or persons, firm, association, company or corporation, shall manufacture, sell, expose, or offer for sale any compound or mixture branded or labeled as and for honey which shall be made up of honey mixed with any other substance or ingredient. There may be printed on the package containing such compound or mixture statement giving the ingredients of which it is made; if honey is one of such ingredients it shall be so stated in the same size type as are the other ingredients, but it shall not be sold, exposed for sale, or offered for sale as honey; nor shall such compound or mixture be branded or labeled with the word "honey" in any form other than as herein provided; nor shall any product in semblance of honey, whether a mixture or not, be sold, exposed, or offered for sale as honey, or branded or labeled with the word "honey" unless such article is pure honey.

The value of such a law in safeguarding the bee-keeper's market and protecting the consumer against fraud can scarcely be estimated. Imitations are still to be had in the market but they sell for just what they are and the consumer who cares to use them buys them at a lower price than he would have to pay if they were permitted to be sold as honey.

Net Weight Labels.—The provision of the law which requires every package to have the net weight measure or numerical count plainly marked on the label necessitates stamping every section of comb honey as well as every jar holding extracted honey with the amount of honey it contains. This provision led to much complaint among small producers at first. After a few months trial it is being demonstrated that it is really an advantage to the comb honey producer who is up to date and has much honey to market. This requirement applies to all interstate shipments which come under national jurisdiction. The paragraph reads as follows:

> A food product will be deemed to be misbranded: If in package form, the quantity of the contents be not plainly and conspicuously marked on the outside of the package in terms of weight, measure, or numerical count; provided, however, that reasonable variations shall be permitted, and tolerances and also exemptions as to small packages shall be established by rules and regulations made in accordance with the provisions of section three of this act.

A similar requirement is made by some State laws so that the net weight must be marked on packages sold to the local trade as well as those shipped to distant markets.

The effect of this provision is to keep much ungraded honey out of competition with a first-class product. The large producer finds it an easy matter to provide cartons on which are printed the weights of the various grades and as each section is graded it is placed in a carton of the proper kind.

QUESTIONS

1. Note the peculiar conditions that surround the bee-keeper in his relation to the public.
2. Discuss the usual causes of trouble between bee-keepers and neighbors.
3. When will bees be regarded as a public nuisance?
4. Discuss the spraying of fruit trees while in bloom.
5. Discuss the laws for control of bee diseases.
6. What is the effect of the laws relating to the adulteration of honey?
7. Summarize briefly the various laws relating to beekeeping.

INDEX

Adulteration of honey, 296
    of wax, 199
Advertising, exhibits for, 269
    general, 268
    methods of, 267
Alexander feeder, 134
    plan of making increase, 104
    strainer, 190
Alfalfa, 60
    region, 49
Alley plan of queen rearing, 124
Apiary, arrangement of, 36–41
Apprenticeship, value of, 18
Artichoke, 63
Ash, source of pollen, 80
Aster, 69

Basswood, 60
Beech, 80
Bee-escapes, 161
Bee-keepers, studious, 9
Bee-keeping, advantages of, 10
Bees as pets, 1
Beeswax, adulteration of, 199
    color of, 197
    melts at low temperature, 196
    production of, 195
    substitutes for, 199
    uses of, 198
    see also Wax
Birch, source of pollen, 80
Bitter honey, 81
Bitterweed, 81
Black bees, 98
Bleaching wax, 204
Blending honey, 265
Boiler press for wax, 204
Bonney advertising stickers, 268
    hive markers, 44
    postcard, 280
Book-keeper, successful bee-keeper, 15
Booklets for advertising, 276
Box-elder, source of pollen, 80
Breeding to produce non-swarming bees, 157
Brood rearing, feeding for, 130
Buckwheat, 61
Bulk honey, 194
Business, bee-keeping as exclusive, 16
Button bush, 78
Buying bees, 27

Cage method of introducing queens, 114
Candied honey, liquefying, 191
    retailing, 193
Candy stores, bees at, 285
Canvassing, to sell honey, 280
Carniolans, 99
Carpenter, a bee-keeper, 13
Catnip, 78
Caucasian bees, 99
Cellar, essentials of good, 254
    for wintering, 252
Cells, care of queen, 126
Chaff hives, 234
Chestnut, source of pollen, 80
Chilled brood, 222
Chunk honey, 194
Clerk, successful bee-keeper, 13
Clipping queens, 101
Closing the season, 162
Clover region, 48
Colonies, to tell strong, 30
Colorado plan of inspection, 295
Comb bucket, 25
    honey, care of, 262
        production, 136
Combs, care of empty, 228
Commercial queen rearing, 123
Commission houses, selling through, 260
Containers for honey, 190
Control of bee diseases, 292
    essentials of, 5
Corn, source of pollen, 80
Cover, for hive, 23
Cranberry, bee as pollenizing agent, 84
Crownbeard, 66

Cucumber, bee as pollenizing agent, 85
Cup-plant, 65
Cutting out queen cells, 102, 158
    prices of honey, 282
Cyprian bees, 98

Dadant method of wintering, 242
    hive for extracted honey, 173
Dandelion, 48, 53
Demaree method of swarm control, 185
Demonstrations, with bees, 5
De-queening during honey flow, 158
Devil's darning needles, 224
Diseases, 206
    American foul brood, 207
    dysentery, 223
    European foul brood, 215
    laws for controlling, 292-296
    sacbrood, 221
    treatment of, 212, 219, 223
Disinfecting, for foul brood, 215
Division board feeder 134
Doolittle feeder, 134
    method of queen rearing, 124
Doorweed, 74
Double-walled hives, 245
Dragon flies, 224
Drone, 95
Dysentery, 223

Elm, source of pollen, 80
Empty combs, care of, 206
Enemies of bees, 206
Entrance feeder, 134
Entrance, width of, 186
Equipment, for beginner, 18, 19
    for comb honey production, 137
    minor, 23
Excluders, 186
Exhibits, at fairs, 269
Experience, getting, 18
Extracted honey, packages for, 263
    power for, 168
    production of, 165
    storage tanks for, 169
Extracting, 188
Extractors, 165

Failures, from lack of experience, 16

Fairs, exhibits at, 269
Fall flowers, 63
Farmer, bee-keeper, 16
Feeding bees, 128
    for reserve supply, 129
    preparation for, 129
    to stimulate brood rearing, 130
Feeders, Alexander, 134
    division board, 134
    entrance, 134
    Miller, 133
    Minnesota, 131
    tin-pan, 133
Fertile workers, 94
Figwort, 78
Flaxboard, 24
Florida, honey plants of, 50
Foul brood, 206
    American, 207
    European, 215
Foundation, full sheets of, 180
    in sections, 145-198
Frames, 174
Fruit bloom, 55, 85
Fumigation for wax moths, 163

German bees, 98
Getting acquainted with bees, 1
Gloves, need of, 20
Goldenrod, 61
Grading, extracted honey, 265
    honey comb, 257
        official rules for 258

Heartsease, 74
Hive, for extracted honey, 173
    kind to adopt, 19
    markers, Bonney, 45
    marks, 43
    observatory, 25
    spacing, 38
    stands, 39
    to open, 5
    tool, 21
Hiving swarm, 103
Hoffman frame, 174
Home markets, 260
Honey flow, of short duration, 9
    house, 175
    method of introducing queens, 117

INDEX

Honey producers, women successful, 11
    pump, 169
    ripening of, 187
    straining, 189
Honey-dew, 50
    unsatisfactory for wintering, 128

Increase, 100
Inspectors, business of, 229
    requirements for successful, 231
Italian bees, 27, 98

Joys of bee-keeping, 7

Knotweed, 74

Labels, honey, 266, 269
    net weight, 298
Lady's thumb, 74
Laws, against poisoning bees, 291
    for control of bee diseases, 292
    net weight, 298
    relating to adulteration of honey, 296
    restricting shipment of bees, 295
    spraying, 291

Maple, 52, 80
Market, comb honey, 137
    home, 260
Marketing, by mail, 276
    by canvassing, 280
    co-operative, 265
Mice, injury from, 224
Milkweed, 75
Miller, A. C., smoke method of introducing queens, 115
Miller, Dr. C. C., dequeening method, 160
    feeder, 133
    method of making increase, 111
    method of queen rearing, 122
    plan of producing comb honey, 154–156
    Smoke method of requeening, 115
Minnesota feeder, 131
Mosquito hawk 224
Mountain laurel, 81
Moving bees, 28

Nectar, sources of, 46
Net weight labels, 298
New York, law for controlling bee diseases, 292
    law for sale of honey, 297
Nuisance, bees as, 286
Number of bees in a colony, 94

Observatory hive, 25
Occupants of the hive, 88
Odor of stable offensive to bees, 6
Orchards, bees as pollenizing agents, 85
Outlook for beekeeping, 17
Overstocking, 82

Packages, fancy, 273
    for comb honey, 261
    for extracted honey, 263
Packing cases for wintering, 247
Packing for winter and summer stands, 244
Palmetto, 77
Paper cases for wintering, 240
Partridge pea, 67
Peddling honey, 280
Pitting bees for winter, 243
Poisoning bees, 291
Poisonous honey, 81
Pollen, sources of early, 52, 79, 80
Pollenizing agents, 84
Portable outfits for extracting, 175
Porter bee escape, 161
Postcards, for advertising, 175
Preparation, advance, 150
Prices, control of, 264
    retail, 274
Prior rights, 83
Protection, of hives in spring, 239

Queen, 88, 92
    cells, cutting, 102, 158
    clipping the, 101
    excluders, 186
    influence of, 236
    introducing, 112, 220
    rearing, 121
    replacing, 27

Races of bees, 98
Receipts for cooking with honey, 277

Retail markets, 274
Returns from beekeeping, 12
Rhododendron, 81
Robber fly, 225
Rosin weed, 65

Sacbrood, 221
Sage, 77
Sale of honey, laws concerning, 296
Saw palmetto, 77
Seasons management, 149
Sections, for comb honey, 138
    removing from super, 163
Sentinels at entrance, 4
Separators, 141
Shade, value of, 36
Shipment, of queens, 296
    restrictions of, 295
Shipping cases, for comb honey, 261
    for extracted honey, 263
Signs, 270, 271
Simpson's honey plant, 78
Situations for keeping bees, 11
Skunks, 224
Smartweed, 74
Smoke, use of, 5
Smoker, 22, 23
Smoker fuel, 22
Snakefeeders, 224
Sneezeweed, 81
Snow-on-the-Mountain, 81
Solar wax extractors, 200
South, honey plants of, 49
Space under brood nest as swarm prevention, 160
Spacing hives, 38
Spiders, 225
Split sections, 144
Spotting clothes, 284
Spraying when in full bloom, 291
Starters, putting in, 147
Starved brood, 222
Starwort, 69
Steam press, 203
Sting, 2
Strong colonies, important, 83
Sunflowers, 63
Super springs, 148

Supers, enticing bees into, 151
    putting on, 184
Supplying empty combs, 110
Swarm control, 156
    Demaree method of, 185
Swarming, 100
Tools for apiary, 20
Transferring, 32, 33, 34
Trembles, caused by boneset, 72
Trouble, causes of, 284
Truck crops, 57

Uncapping boxes, 170
    knives, 173

Veil, 20, 21
Ventilation, of hives, 30, 186

Walnut, source of pollen, 80
Watering devices, 41, 42
    places, bees at, 284
Water, method of introducing queens, 117
    need of, 41, 238
Wax, adulteration of, 199
    bleaching, 204
    cooling, 205
    moths, 225
        fumigation for, 163
    press, 203
    production of, 195
    rendering, 200
    substitutes for, 190
    uses of, 198
Weak colonies, care of, 154
White snakeroot, 70
Whitewood, 60
Wild bergamot, 74
Wild cucumber, 79
Willow, 52, 74
Wintering, avoid failing queens in 236
    best feed for, 235
    essentials of successful, 234
    influence of the queen in, 236
    methods of, 239-256
    protection from winds, 237
Wiring frames, 181
Worker bees, 92

www.ingramcontent.com/pod-product-compliance
Lightning Source LLC
Chambersburg PA
CBHW062212220526
45471CB00009B/3167